CONTENTS

ABSTRACT

This investigation has been mainly confined to providing descriptions for the identification of the various winged morphs of the adelgid species occurring in Britain. Morphological keys and descriptions, together with 32 figures, are given for the determination of the 14 winged morphs of the 9 species having alatae. In addition, a key to the adult apterae of the genus *Pineus* is included so as to enable identification to be made of those species which do not have a winged form. Only in the cases of *Adelges viridis* and *A. abietis* gallicolae, and *Pineus orientalis* and *P. pini* sexuparae, is it still necessary to depend upon life cycle information for separation down to species. Three of the species described, *A. viridana*, *P. pineoides* and *P. orientalis*, have not previously been recognised in Britain. Details are given of transfer experiments carried out to check host alternation pattern in the alatae of certain species. A world check list of adelgids, their common hosts and distribution is appended.

ACKNOWLEDGMENTS

I wish to express my gratitude to Mr. D. Bevan for his encouragement and helpful criticism of this work, and to Dr. V. F. Eastop of the British Museum for his advice on taxonomic treatment. I would like to thank the Forestry Commission for sponsoring the publication of this work by H.M. Stationery Office. It consists of an extract of a thesis accepted by the University of London for the degree of Master of Science. My thanks are also due to Mr. L. C. W. Bonacina for reading the manuscript; to Mr. N. R. Maslen for assistance in the translation of the German articles; and to the Forestry Commission's Photographic, Library and Workshop staff at Alice Holt Research Station for their help in various ways.

FORESTRY COMMISSION BULLETIN

No. 42

3/02

Conifer Woolly Aphids (Adelgidae) in Britain

By C. I. CARTER, M.Sc.

Forestry Commission

LONDON: HER MAJESTY'S STATIONERY OFFICE

1971

A

SBN 11 710134 6

Chapter 1

INTRODUCTION

A. GENERAL

Insects of the family Adelgidae which feed on conifers can be readily placed in one of the two genera *Pineus* or *Adelges* on the basis of their spiracular arrangement. Determination at the specific level, however, is in some stages difficult or even impossible, in spite of the fact that only a dozen or so species have been recorded in Europe. Most authors, including Börner and Heinze (1957) and their contemporaries, have found that the first instar is the only stage to provide reliable and constant morphological characters suitable for specific determinations within the Adelgidae. Determinations are often required of other stages, and in particular it would be useful to be able to identify alatae. Separation of these winged adults on morphological grounds is simple only in the case of one or two species. It therefore follows that winged individuals have been "identified" in the past by their association with a particular species of conifer.

B. PREVIOUS WORK

The more intensive early taxonomic and biological studies of this family were made in Russia and Germany by Cholodkovsky (1896) and Börner (1908). They compiled much of the available information on the European species and gave valuable illustrations to some of the forms. In England Burdon (1908a and 1908b) reviewed much of Cholodkovsky's work on the biology of adelgids for the purpose of controlling certain species on larch in Britain. Marchal (1913) working in France, contributed further biological and morphological observations on the Silver fir and pine feeding species.

In Britain, most of the investigations have been confined to studies on individual species. Steven (1917) and Speyer (1919 and 1923) reviewed the European work and made further contributions on the life cycles of the larch adelgids. Chrystal (1925) has made a detailed study on the method of feeding of the Silver fir adelgids and Varty (1956) has given some descriptions of the forms of two species and their biology in Scotland. Chrystal (1922) and Cameron (1936) made intensive studies of the life cycles and morphs of *Adelges cooleyi* occurring in Britain.

Pschorn-Walcher and Zwölfer (1958) have reviewed the classification of the species and forms of the Silver fir adelgids in Europe. Their classification is based mainly on morphological features of the first instar nymphs. Eichhorn's (1958) chromatographic and morphological investigations gave some confirmation to the existence of three main species commonly occurring on Silver fir in Europe. Similar morphological studies on the first instars of this group were made in Scotland by Busby (1962).

The larch-spruce species, *Adelges viridis*, has recently been examined again by Steffan (1962) in Central Europe, and apart from confirming the existence of the separate but related *A. abietis*, revealed a third species which he called *A. segregis*. He only gives morphological features to identify these three species at the first instar stage.

Börner published a number of papers on the systematics of the adelgids following his monograph in 1908. The eight genera that Börner recognised appear in the keys given by Börner and Heinze (1957), and are accompanied by short descriptions on most of the species known up to that date. Keys to the European species have since been given in greater detail by Heinze (1962), but as with Börner's later work, much emphasis is given to using the first instars.

Pasek (1954) has illustrated the adults of some of the adelgids occurring in Czechoslovakia. Comparative descriptions and comprehensive keys to the adult winged forms of the European species appear to be absent from all recent literature.

Descriptions of the American adelgids are given in a monograph by Annand (1928). He also has a key to the apterae and alatae which is of some value, as several of the species described occur in Britain. Inouye (1953) has figured some of the alate forms of Japanese adelgids. Some of these species appear to be the same, or races of, the European species.

C. OBJECTIVES

This publication embodies the results of some studies made between 1965 and 1967 on adelgids occurring in Britain. The investigation was stimulated by the sparse literature relating to the identification and flight activity of the winged forms. The objectives were as follows:

(1) To adequately illustrate and describe the winged forms of the species occurring in Britain so that they can be identified when collected out of context with the host plant or colonies. This became necessary when requiring to identify winged migrating forms.

(2) To attempt to determine the number of species present.

(3) To investigate the occurrence of the different life cycle patterns in Britain.

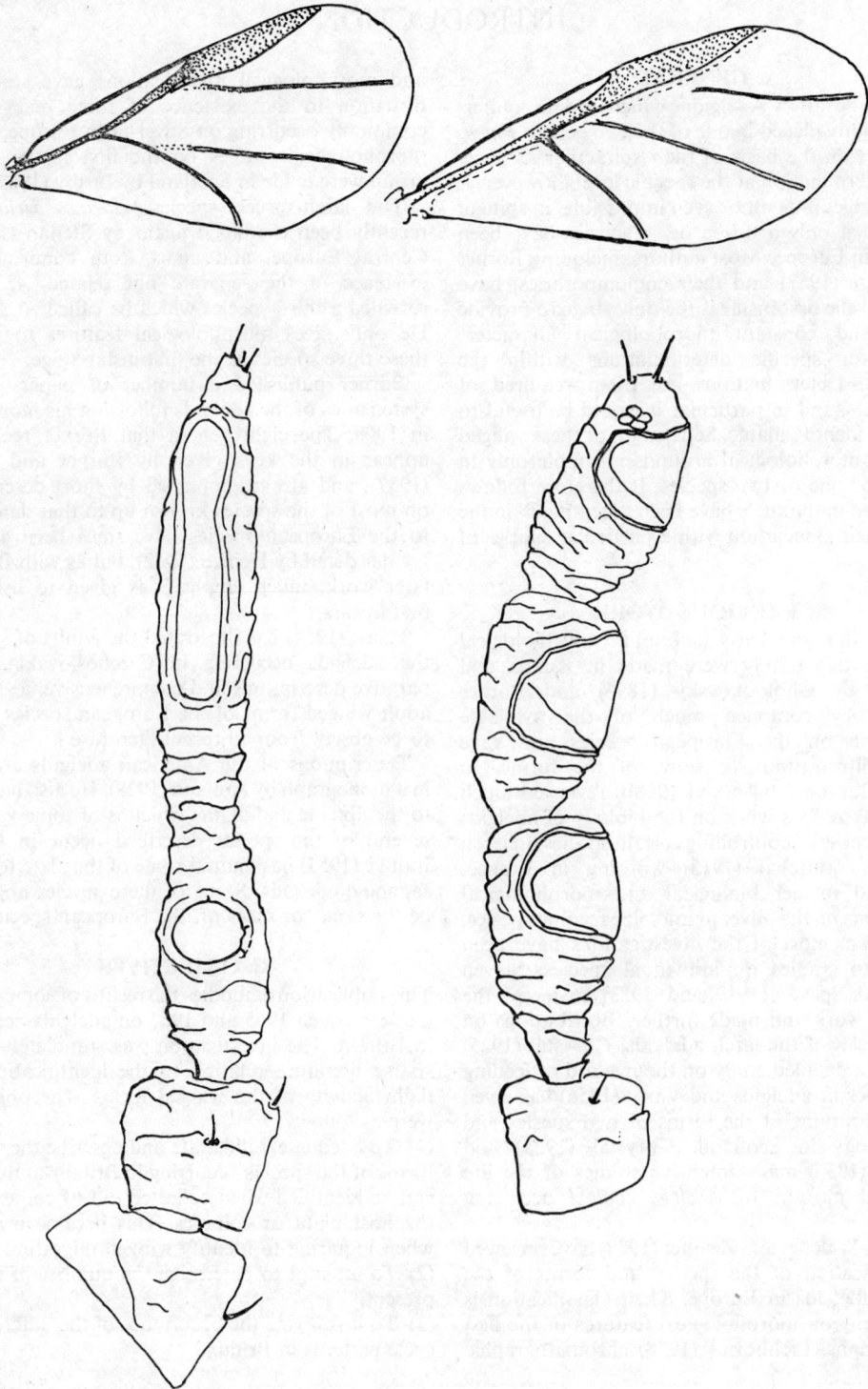

Fig. 1 Forewing and terminal antennal segments of an alate adelgid (right) and of an alate phylloxerid (left).

Chapter 2

SYSTEMATIC POSITION AND GENERALISED LIFE CYCLES OF THE ADELGIDS

A. SYSTEMATIC POSITION

The conifer-feeding insects known as adelgids belong to the super-family Aphidoidea in the order Homoptera. In general, the features of adelgids and aphids are similar in that they are both soft-bodied, have paired membranous wings and feed on plant sap.

Adelgids differ morphologically from true aphids in having short antennal segments, a reduced wing venation, often a highly glandular body surface, in having no siphunculi, and in all the female forms being oviparous. Thus the adelgids are placed in a separate group, the family Adelgidae.

The phylloxerids are also contained in the super-family Aphidoidea as a third family, the Phylloxeridae. The winged phylloxerids superficially resemble winged adelgids, and the female forms of both families are oviparous, hence their inclusion in the same family in earlier work (Annand 1928). They differ morphologically from one another in their form of wings and antennae. The cubital and anal veins of the forewings arise independently from the radial sector in the Adelgidae whereas in the Phylloxeridae two cubital veins occur having a common origin. In the Adelgidae, the antennae of alatae consist of five segments and normally have three primary rhinaria; but in the Phylloxeridae there are only three segments and two rhinaria (see fig. 1). All of the phylloxerids feed exclusively on dicotyledonous plants whilst all the adelgids feed exclusively on the sap of conifers (Pinaceae).

B. LIFE CYCLES

The life cycles of species within the Adelgidae are variable and complex. The basic pattern is a pentamorphic holocycle which takes two years to complete. All but one of the female forms are parthenogenetic. Two different host plants are involved in the holocycle, a primary host which is always a spruce (*Picea* sp.) and a secondary host which is always a conifer of another genus and may be a silver fir (*Abies*), a larch (*Larix*) a pine (*Pinus*), a Douglas fir (*Pseudotsuga*) or a hemlock (*Tsuga*). Migrations between the primary host to the secondary conifer host takes place by winged (alate) forms.

HOLOCYCLIC LIFE CYCLE (see plate 11 in central inset)

The sequence of forms in the complete life cycle is as folows:

(1) The apterous sexual forms are short-lived and very small and occur only on the prlmary host (*Picea* spp.) in the summer. Each fertilised female of this sexuales generation lays a single egg which hatches within a few days to give the fundatrix.

(2) The fundatrix nymph settles down on the current year's shoot, frequently near to a bud. The fundatrix grows very little at first, usually moulting once before the winter. Its renewed feeding activity in late March of the new year seems closely synchronised with the flow of sap, as by the time the buds have started to swell in late April the fundatrix is a mature female. On reaching maturity the fundatrix lays, over a period of a few days, a large cluster of eggs which hatch to give the gallicolae.

(3) The gallicola nymphs hatching out of these eggs coincide in early May with the loosening of the bud scales so that there is fairly easy access for these active nymphs to the tender young shoot in the bud. This is the stage which is associated with the production of the characteristic galls of the Adelgids. [It has been suggested by Plumb (1953) that the feeding activity of the fundatrix in the spring causes the presumptive needle and stem tissues in the opening spruce bud to be deformed, resulting in a cone-like gall. The gallicola nymphs may also contribute to the deformity as they crawl into the bud. While they feed on the sap of the young tissue the gall forms round them and they become imprisoned. The tissues of the gall eventually ripen at various times during the summer, according to the weather and species. In so doing the deformed needle bases which form the bulk of the gall become separated, releasing the gallicolae]. The gallicolae emerge from the gall and crawl up on to the needles. These are now winged nymphs, so that after an ecdysis they can expand and harden their wings ready for flight. Migration of the gallicolae takes place throughout the summer and is to the secondary host in holocyclic species. On migration the gallicolae settle on needles and very soon lay their eggs on the needle under the protection of their folded wings. These hatch to give sistens, sometimes referred to as exsules.

(4) This winter sistens (or hiemosistens) remains in the first or second instar until the spring. Like the fundatrix, the winter sistens becomes mature in the spring and lays large clusters of eggs. These eggs hatch to give progrediens.

(5) The progrediens nymphs usually develop simultaneously into two types of morph, the progrediens apterae (5b), which are wingless, and the sexupurae (or progrediens alatae 5a) which are winged.

(5a) The sexuparae of holocyclic species migrate back to the new needles of the primary (*Picea* sp.) host in May and June. Eggs are laid on the needle

under their wings in the same manner as the galli-colae. The eggs of the sexuparae hatch out as males and females, the sexuales generation.

(5b) The apterous progrediens may produce several further generations of progrediens in a favourable summer, resulting in a sistens generation which then overwinters; there may also be a summer sistens form (aestivosistens). The winter sistens

forms so produced are morphologically identical to those derived direct from the gallicolae and similarly give rise to sexuales and progrediens, Speyer's (1923) observations on *Adelges laricis* Vallot indicate that the older the colony on the secondary host plant, the larger the proportion of sexuparae produced.

For ease of reference this basic pentamorphic holocycle is reproduced in tabular form.

THE SEQUENCE OF FORMS IN AN ADELGID PENTAMORPHIC HOLOCYCLE

Season	Primary Host PICEA Species	Secondary Host ABIES, LARIX, PINUS PSEUDOTSUGA, TSUGA
Summer	(1) SEXUALES Small wingless males and females late May to July.	
Winter	(2) FUNDATRICES Nymphs from August to April, apterous adults in May.	
Summer	(3) GALLICOLAE Nymphs from May to August, winged adults from June to September according to species.	
	Migration ———————————→	then lay eggs
Winter		(4) SISTENS (or Exules) as nymphs from June to April, apterous adults by May.
Summer	then lay eggs ←——————————— Migration	(5) PROGREDIENS (2 types) (a) SEXUPARAE winged adults May to July. (b) PROGREDIENS apterous adults May to August, may be several generations of this form.
	(1) SEXUALES (as above)	(4) SISTENS (as above)

ANHOLOCYCLIC LIFE CYCLE

Adelgids that do not alternate between *Picea* and another conifer host are sometimes termed 'monoecious' (Steffan 1962). Such species do not have all of the morphs in the above table represented in their life cycle, i.e. they are anholocyclic. This reduction is in all cases accompanied by the loss of the gamic morphs, so that reproduction only takes place by parthenogenesis. Anholocyclic adelgids may be confined to any one of the conifer genera. There are four types of anholocycly occurring in Britain. These are as follows:

(i) The insect exists only on *Picea* and produces two generations a year. The morphs have the appearance of stages 2 and 3 above. As there is no sexuales generation, stage 2 is referred to as the

pseudo-fundatrix. (Example: *Adelges abietis*).

(ii) Another type only has apterous forms on *Picea* but there are no galls formed. The name pseudohiemosistens has been used by Steffan (1963), as the apterous forms do not follow the gallicolae, nor do they occur on the secondary host. (Example: *Pineus pineoides*).

(iii) Those species occurring on the secondary host that have all the stages in 4 and 5 above. They may produce sexuparae, but are incapable of producing males. (Example: *Pineus pini*).

(iv) In a similar way, other species have stages 4 and 5 on the secondary host and produce alatae, but these are not sexuparae as they give rise to parthenogenetic females resembling the sistens. (Examples: *Adelges piceae* and *A. viridana*).

Chapter 3

KEYS AND DESCRIPTIONS OF THE BRITISH ADELGIDAE

A. GENERAL MORPHOLOGY OF THE FORMS DESCRIBED

THE FIRST INSTAR NYMPHS

These are very small, less than 0.5 mm long. The head, thorax and abdomen are fused to form an oval shape. Small, three-segmented, antennae typically have a long third-segment with a long apical seta. Legs and mouthparts are complete, but the eyes are represented only by three ocelli on either side of the head region.

In order to see any finer morphological details it is essential that good microscopical preparations are made. Certain features on the dorsal surface can be seen much more easily with a phase-contrast microscope, especially when the legs and ventral parts have been dissected away.

The dorsal surfaces of first instar nymphs have a regular arrangement of segmental chitinous plates (see fig. 7). Their relative sizes, arrangement and ornamentation with wax glands are of considerable taxonomic use. There are basically six longitudinal rows of dorsal plates. Conforming to the terminology of Annand (1928) the middle two rows are called mesial plates; next to these on either side are the pleurals and outside the pleurals are the marginals. They are segmentally arranged on the abdomen but on the thorax several plates may be fused to form a shield (see fig. 7). Similarly the head segments may have the plates fused but when wax gland facets are grouped together on chitinous shields they often show the basic segmental pattern. Two pairs of large lateral spiracles are evident in the thorax, but the paired

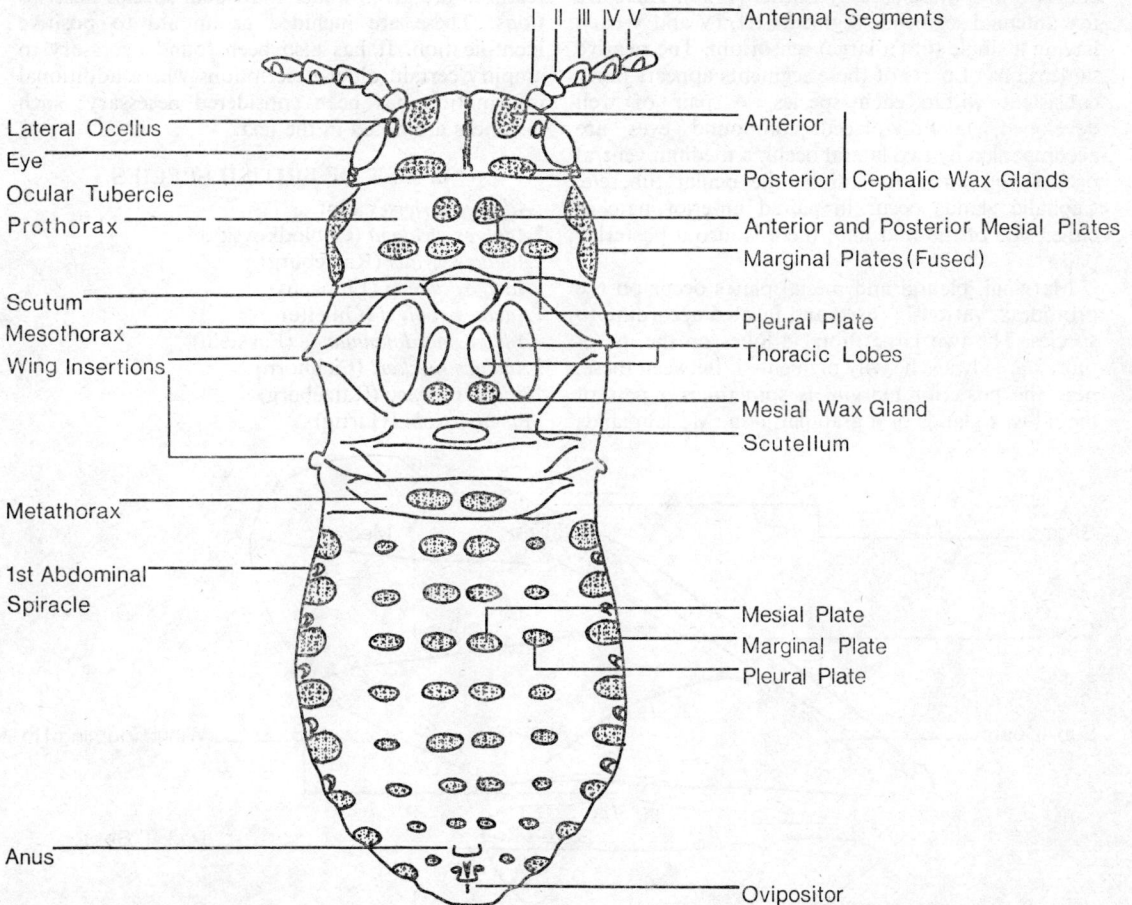

Labels on diagram:
- I II III IV V — Antennal Segments
- Lateral Ocellus
- Eye
- Ocular Tubercle
- Prothorax
- Scutum
- Mesothorax
- Wing Insertions
- Metathorax
- 1st Abdominal
- Spiracle
- Anus
- Anterior
- Posterior } Cephalic Wax Glands
- Anterior and Posterior Mesial Plates
- Marginal Plates (Fused)
- Pleural Plate
- Thoracic Lobes
- Mesial Wax Gland
- Scutellum
- Mesial Plate
- Marginal Plate
- Pleural Plate
- Ovipositor

Fig. 2 Diagram of the dorsal surface of an alate adelgid. The basic pattern of wax glands is shown stippled.

abdominal spiracles are difficult to detect until the insect has reached the second instar.

The apterous winter (sistens) forms are more heavily chitinised than the summer forms and are therefore much easier to examine and identify.

ADULT PARTHENOGENETIC APTEROUS FEMALES

These have the same overall basic chitinous plate arrangement as the first instars, but in the genus *Pineus* there is greater fusion of plates in the anterior region. The adult females are much larger, up to about 2 mm long, and show a general increase in the number of wax gland facets. There is always an ovipositor present.

ALATE FEMALES

These resemble the general aphidian form, having a distinct head, thorax and abdomen (see fig. 2). The morphological terms used in the alate descriptions conform with those used by Cottier (1953). There are five antennal segments, segments III, IV and V each having a single (often large) sensorium. The relative dimensions of parts of these segments appears fairly consistent within each species. A pair of well developed, laterally-placed compound eyes are accompanied by two lateral ocelli, a medium ventral ocellus and three ocellani on the ocular tubercle. Cephalic glands occur in paired anterior patches either side of the mid line; there is also a posterior pair.

Marginal, pleural and mesial plates occur on the prothorax, variously fused and faceted according to species. The two large thoracic lobes on the mesothorax are always heavily pigmented. Between these, near the posterior margin, is sometimes a pair of mesial wax glands or a granular area. Mesial glands

on the metathorax are frequently large oval areas. Mesial, pleural and marginal rows of plates are usually represented on the abdominal segments by wax gland areas. The marginal rows are always present, but the pleural rows are frequently reduced or absent on the anterior and posterior segments. The mesial rows, when present, decrease in size posteriorly. An ovipositor is also present, being similar to the type possessed by apterae.

The shape and venation of the wings is a useful character in segregating certain species and even forms (see fig. 3). In the forewing the stigma of the subcosta varies a little in position; the anal and cubital veins are sometimes distinctly curved or straight. In the hind wing the media may be absent or arise at various angles to the radial sector, this also may be curved or straight according to species.

Descriptions and relevant life cycle data of hitherto undescribed morphs recorded during the research are given under individual species descriptions. These are included as an aid to positive identification. It has also been found necessary to amplify certain older descriptions where additional information has been considered necessary; such instances are noted in the text.

B. LIST OF BRITISH SPECIES

Adelges laricis Vallot
*Adelges viridana (Cholodkovsky)
Adelges viridis (Ratzeburg)
Adelges abietis (Linnaeus)
Adelges cooleyi (Gillette)
Adelges nordmannianae (Eckstein)
†*Adelges merkeri* (Eichhorn)
†*Adelges piceae* (Ratzeburg)
Pineus strobi (Hartig)

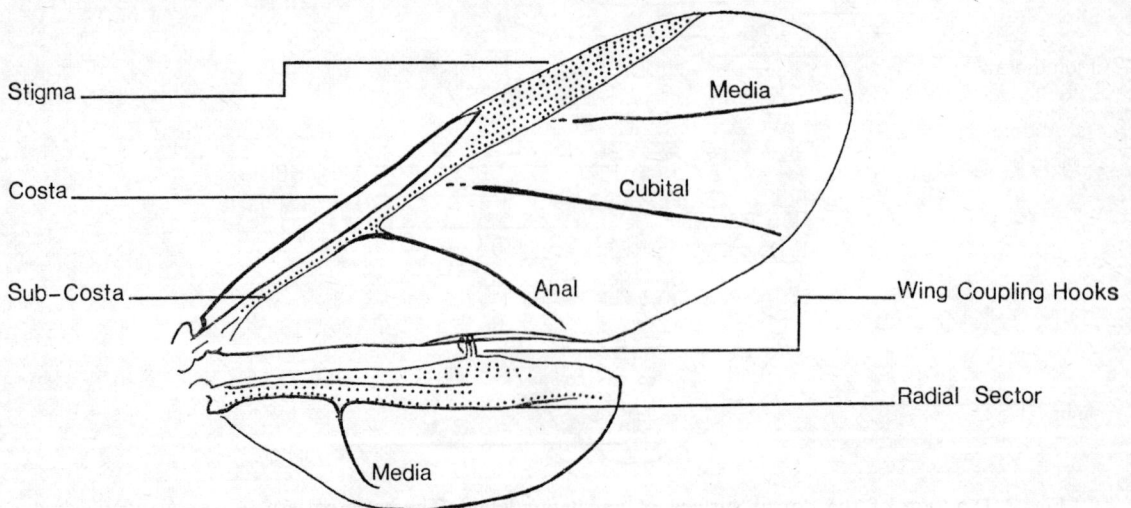

Fig. 3 Basic wing venation of an adelgid.

Pineus pineoides (Cholodkovsky)
Pineus pini (Macquart)
Pineus orientalis (Dreyfus)

The species marked * have been recorded as new to Britain during the course of this work (Carter 1969).

With the exception of the two species marked †, all of the above species are included in the keys and descriptions that follow. Winged forms of these two species have not been recorded under natural conditions in Britain. Furthermore Varty (1953) who produced winged forms of *A. piceae* under experimental conditions, concluded that the occasions when this could occur naturally in Britain would be rare, this being due to the necessary phenological synchronisation of the insect with the host plant. *A. merkeri* has only been recorded once in Britain, by Busby (1962), having been identified only from the first instar sistens occurring on *Abies*.

C. KEY TO THE GENERA OF THE FAMILY ADELGIDAE

ADULT FEMALES POSSESSING AN OVIPOSITOR
1 Adult morphs, having four distinct pairs of lateral abdominal spiracles . *PINEUS* Shimer
– Adult morphs, having five distinct pairs of lateral abdominal spiracles . *ADELGES* Vallot

ALATAE
1 Antennal segments III and IV with broad and square-ended distal ends, rapidly tapering proximally below their sensoria to small, distinct articulating surfaces. Antennal sensorium on IV occupying approximately the distal half of the segment and extending more than half way round it. Antennal segment V not appreciably longer than III or IV. Marginal and mesial wax gland areas on abdomen large and oval, facets clearly visible. 4 distinct pairs of lateral abdominal spiracles between the marginal wax-gland areas . . . . *PINEUS* Shimer
– Sensoria of antennal segments III and/or IV occupying less than half the length of the segment. Antennal segment V appreciably longer than III or IV in some species. Gradual taper of antennal segments IV and V proximally to the position of the sometimes indistinct articulating surfaces. If an angular taper on IV and V then they possess small slit like sensoria. 5 distinct pairs of lateral abdominal spiracles between the marginal wax-gland areas . *ADELGES* Vallot

D. Genus *ADELGES* Vallot 1836

In all species of this genus there are five pairs of lateral abdominal spiracles present, but sometimes these are very indistinct in the first instar stages. All stages have an elongate form. There is much variation in chitinisation and wax gland distribution

but both characters are constant for one morph of any one species. The last instar sexupara nymphs have head and prothoracic plates which are separated.

Börner published a number of contributions, commencing with his monograph in 1908, in which the genus *Chermes* was divided up into seven genera. The name *Chermes* has now been suppressed because of confusion (Eastop 1966) and *Adelges* used in its place. Börner and Heinze (1957) have divided the genus *Adelges* up into six parts, *Adelges* Vallot, 1836; *Sacchiphantes* Curtis, 1844; *Cholodkovskya* Börner, 1909; *Gilletteela* Börner, 1930; *Dreyfusia* Börner, 1908; and *Aphrastasia* Börner, 1908. Annand (1928) seemed strongly opposed to Börner's division of the genus *Adelges* into six genera, based as it is on morphological features which are so subtle that in another insect group they would only be recognised as specific differences. Annand's (1928) opinion that there are only two genera in the family has been accepted during this work.

In the World list of species in the Appendix, Börner's genera therefore appear as sub-genera.

It is interesting, however, that in most cases, differences have been found to be more distinct between the holocyclic species in the first instar of the sistens than in any adult apterous stage. This was also found to be true in the case of the alatae in this study. It seems that speciation has taken place more at the first instar level of the sistens than the winged adults. One explanation for this could be that first instar sistens forms of the various genera hibernate on a particular part and species of secondary host plant, whilst the winged adults only feed on new shoots of the secondary host or are produced in galls (modified new shoots) on *Picea*.

Many species are holocyclic, having a two yearly pentamorphic life cycle as given earlier in Chapter 1; the primary host being *Picea* sp.; the secondary hosts being a species of one of the following conifer genera – *Abies, Larix, Pseudotsuga, Tsuga*. Some species are much more abundant on the secondary host where several sistens and progrediens generations may occur over a number of years without obligatory migrations intervening. Consequently very large colonies may result. The feeding sites on the secondary host vary between bud bases, shoots, needles and stems, certain plants show striking effects as a result of their sap sucking activity.

Several species have been reported in the past as anholocyclic (Chrystal 1922) living either on the primary or secondary host. Some of these have almost identical morphology with the corresponding stages of holocyclic species and are only separable on biological grounds or by using statistical morphometric techniques (Philiptshenko 1916).

Cone-like galls are produced by the feeding activity of the fundatrix and her progeny on *Picea* shoots. Galls tend to be of the same pattern for an adelgid species on the particular species of *Picea*. Usually the developing nymphs within the gall are in separate chambers, and they cannot get out to moult and fly away until the gall tissue starts to dry and shrink.

Seven species of this genus have been identified as occurring in Britain during the course of this work. One of them, *Adelges viridana* (Cholodkovsky), has not previously been recorded in Britain. Only six of these species normally have winged forms in Britain, these appear in the keys and descriptions below.

Key to the British Species in the Genus Adelges

ALATAE

1   Head and prothorax with wax gland areas having distinct facets with clear lumina. Paired mesial wax gland areas present between mesothoracic lobes. Mesial abdominal glands present at least on segments 1 to 3   2

−   Head and prothorax wax gland areas granular or not having distinct facets and clear lumina. Mesial wax gland areas between mesothoracic lobes absent or very weak and granular without facets. Mesial abdominal wax gland areas not present in distinct pairs   6

2   Paired anterior and posterior cephalic gland areas united to form two continuous longitudinal areas either side of the median epicranial suture. Anterior prothoracic glands on humps with few or no gland facets. Antennal segments III to V with sensoria occupying at least 2/5 of their length. Antennal segment III longer than IV. Antennal IV tapered above and below the sensorium. Marginal abdominal wax gland areas oval; pleural areas weakly developed posteriorly. Cubital of forewing straight. Hindwing media arising at almost right angles from the radial sector and curving distally away from the thorax .   .   .   .
    .   (progrediens alata) *viridana* (Cholodkovsky)

−   Paired anterior and posterior cephalic gland areas normally separate. Gland facets on prothoracic mesial humps variable in number but have distinct facets. Antennal segment III with sensoria not occupying more than 1/3rd of the length. Cubital vein of forewing almost straight or curved towards the distal tip. Hind-wing media indistinct or arising at less than 90° from the radial sector and curving out distally from the thorax   3

3   Antennal segments IV and V with very broad articulating surfaces having diameters at least 2/5 of widest part of the segment. Length of sensoria

on IV and V 2/5 to ½ that of the segments. The sensorium on V is large and has a sinuate margin; the sensorium on III is much smaller and angular. The cubital vein of the forewing slightly curves away from the thorax. Media of the hind wing leaving the radial sector at slightly less than a right angle. The anterior mesial abdominal wax gland areas narrow and angular, the facets of these glands are circular, and on average, about ½ or more their own diameter apart with light or no granulations to their lumina .   .   4

−   Antennal segments IV and V constricted proximally to a narrow articulating surface. Sensoria narrow, only appearing as a slit, always at the distal end, and rarely occupying more than ¼ of the length of each segment. Cubital vein of forewing straight or with a slight sigmoidal curve. Media of hindwing curving out distally from the thorax, forming an acute angle with the radial sector. Marginal wax gland areas of the abdomen large, giving a corrugated profile. Anterior mesial wax gland areas well developed with large facets. The facets are rounded, not truly circular, and are of various sizes. On average the facets are 1/3 or less their own diameter apart .   .   5

4   Antennal segments IV and V tapered below sensorium (dorsal or ventral view). Mesial wax gland areas on the mesothoracic lobes are small and circular, having up to 60 facets. Anterior mesial and anterior marginal wax gland plates on the abdomen oval or elliptic. Hindwing media faint .   .   .   (sexupara) *laricis* Vallot

−   Antennal segments IV and V appear almost parallel-sided below the sensorium (dorsal or ventral view). Mesial wax gland areas large and oval, extending obliquely posterior to the lobes and have 60 or more facets. On the abdomen the anterior mesial wax gland plates are oblong and extend transversely; the anterior marginal plates are crescent or transversely shaped. Hindwing media distinct .   (gallicola) *laricis* Vallot

5   Sensoria on antennal segments III and IV extend less than ½ way round their segments. The wax gland plates on the abdomen in the anterior mesial and marginal position are large and oval. Forewing length 0.44 to 0.55 mm .   .   .
    .   .   .   .   (sexupara) *cooleyi* (Gillette)

−   Sensoria on antennal segments III and IV extend more than half-way round the segments. The wax gland plates on the abdomen in the anterior mesial and marginal position are elongated transversely, the mesial pairs frequently are fused. Forewing length 0.69 to 0.9 mm .   .   .   .   .
    .   .   .   .   (gallicola) *cooleyi* (Gillette)

6   Costa of forewing arched, giving a broad oval shaped wing (see fig. 13). Head wax gland areas absent, or, if present, they are small and very

indistinct, having a granular appearance (see fig. 11). Mesial prothoracic wax gland areas replaced by a small lenticular pigmented transverse plate with undulating ridges (see fig. 11). Large adelgids, 1.4 to 2.2 mm long   .   .   7

– Costa of forewing almost straight (see fig. 22). Head wax gland areas are pale and without facets. Mesial prothoracic wax gland areas represented by a large transverse plate that extends anteriorly (see fig. 20). Small adelgids 0.8 to 1.6 mm long   8

7 Head wax gland areas absent. Antennal segment III parallel sided below sensorium and rapidly constricting at the basal articulating surface. Sensorium on III rounded and smaller than IV. Line of pigment crossing anal and meidal veins undulating   .   .   .   .   .   .
(gallicola) *abietis* (Linnaeus) or *viridis* (Ratzeburg)

– Granular wax gland areas present on the head. Rounded sensoria on III and IV are equal in size. A line of pigment running across the anal and medial veins is parallel to the sub-costa   .   .   .   (sexupara) *viridis* (Ratzeburg)

8 Antennal segment III with rounded sensorium, only 1/3 the length of, and not exceeding half-way round the segment. Sensorium on IV approximately half the length of the segment. Media on hindwing absent   .   .   .   .
.   .   (sexupara) *nordmannianae* (Eckstein)

– Antennal segment III with sensorium almost ½ the length of, and extending half-way or more round the segment. Sensorium on IV half to 3/5 the length of the segment. Media on hindwing present   .   (gallicola) *nordmannianae* (Eckstein)

Descriptions of Species

Adelges laricis Vallot, 1836

This is perhaps the commonest of the three larch-feeding adelgids in Britain and the abundance of waxwool produced on the needles makes it one of the most striking species. Light needle damage can sometimes be mistaken for that caused by *A. viridis*, but morphologically these insects are quite distinct. *A. laricis* was one of the first adelgids to be studied in detail in Britain by Burdon (1908). Speyer (1923) paid particular attention to the pentamorphic life cycle and the reproductive capacity of the parthenogenetic females. Börner (1908) has given good morphological details of some of the forms and Annand (1928) has given a partial description of American gallicolae. These descriptions, however, do not appear to be adequate for separating the winged forms from certain other species.

Fig. 4 Dorsal body surface of *Adelges laricis* (sexupara).

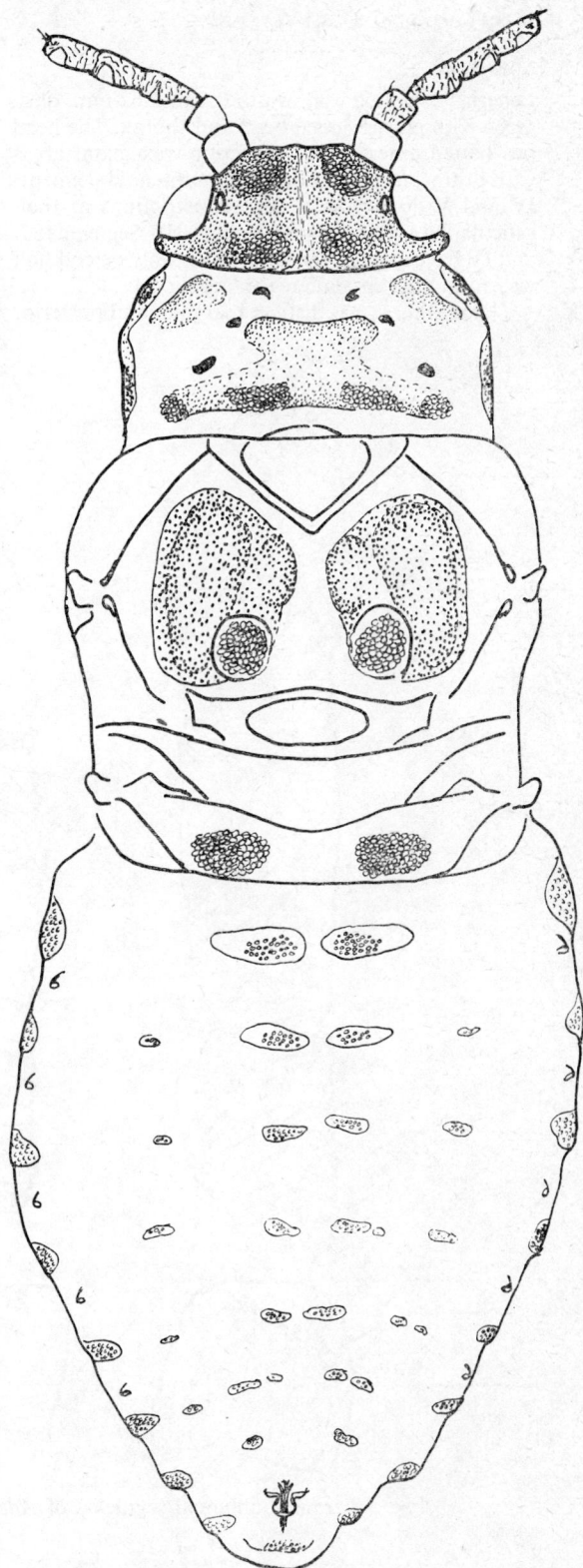

MORPHOLOGICAL OBSERVATIONS

Sexuparae

Length 1.1 to 1.55 mm, width 0.5 to 0.62 mm, dark green with greyish green head and thorax. The head has paired anterior and posterior wax gland areas with distinct facets (see fig. 4). The antennal segments IV and V do not have deep constrictions at their articulating surfaces as in segment III. Segments III and IV have angular sensoria which only extend half way round the circumference (see fig. 5).

The prothorax has distinct marginal and posterior mesial wax gland facets. The pleural and anterior mesial areas are sometimes without facets. Two small, yet distinct, circular wax gland areas occur mesially between the mesothoracic lobes. Behind these is a pair of larger mesial glands on the metathorax.

The forewings measure between 0.47 and 0.55 mm. The medial and cubital veins curve out distally; the anal vein curves back towards the body (see fig. 6). The length of the hindwings varies between 0.25 and 0.33 mm. A rather indistinct media arises at an angle on the radial sector and curves out distally.

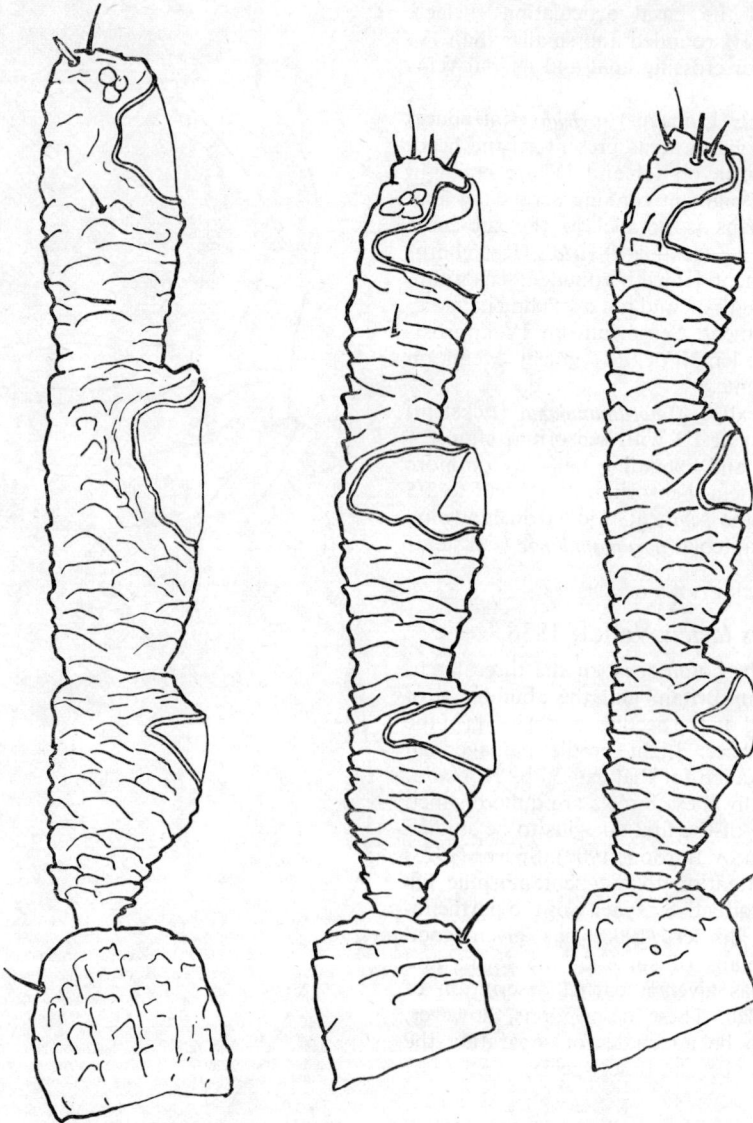

Fig. 5 Terminal antennal segments of *Adelges laricis* (left) and sexupara (right pair).

Distinct wax gland plates are well developed anteriorly in the mesial position, but these decrease in size posteriorly. The pleural plates are almost absent, but the marginal plates are large and oval, becoming smaller in area posteriorly. The wax gland facets of these plates are circular with clear lumina. They are distinctly separate from one another by half or more their own diameter (see fig. 6).

Gallicolae

Length 1.9 – 2.00 mm long, width 0.87 – 0.95 mm. The head has distinct facets on the anterior and posterior wax gland areas. The antennal segments IV and V do not have deep constrictions at the articulating surfaces, so that segments III to V appear fused (see fig. 5). The sensoria on IV and V are longer than those of the sexuparae; the sensorium on segment III is narrow and angular. Each sensorium extends at least half way round the circumference of its segment (see fig. 5).

Wax gland facets in the marginal thoracic position are of various sizes; those on the posterior mesial plates are more uniform. The anterior mesial areas are normally without facets. Between the meso-

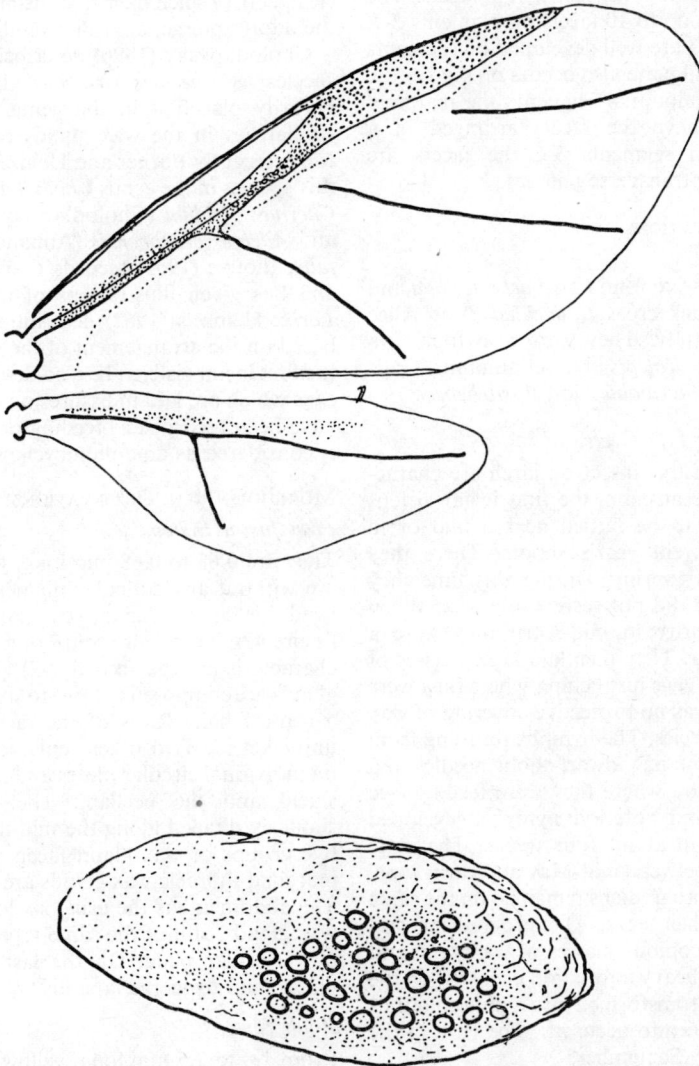

Fig. 6 Wings of *Adelges laricis* (gallicola and sexupara) above, and the mesial abdominal wax gland plate of the sexupara (below).

thoracic lobes are paired wax gland areas, larger in area than those occurring in the sexuparae, and consequently they have many more facets.

The forewings are 0.6 to 0.75 mm long and have the same venation arrangement as the sexuparae. The hind wings are 0.38 to 0.41 mm long; the media is in the same position as in the sexuparae, but here it is much more distinct.

Individual facet size and shape on the abdominal plates are similar to those described for the sexuparae (see fig. 6). Shapes of the glandular areas, however, differ. In the anterior mesial position the plates extend transversely. On segments 4-6 they occur as small irregular patches with just a few facets. The pleural glands have up to 10 facets on segments 2-7. The marginal plates are well developed on segments 1-8, and a single oval plate also occurs on segment 9. On the first two abdominal segments the marginal plates sometimes have the facets arranged in a crescent shape; on segments 3-6 the facets are arranged in oval or transverse patches.

BIOLOGICAL OBSERVATIONS

Host Plants
Heavy infestations were found on *Larix decidua* and *Larix* x *pendula* (back cross *L. decidua* ?) at Alice Holt during this study. They were also found on *L. laricina* and *L.* x *eurolepis* in lesser amounts. Galls are to be found on *Picea abies* and *P. sitchensis*.

Observations on the Life Cycle
The feeding sites of this insect on larch are characteristic. During the autumn, the first instar sistens nymphs were seen to be settled near a bud or in crevices of the current year's shoots. There they remained until early spring. During this time they appeared black and did not secrete any wax. When they reached maturity in mid-April they were a blackish grey colour. They then laid large masses of grey or pale orange eggs just behind where they were feeding, but there was no protective covering of wax as in most other species. The nymphs resulting from these eggs fed on the new dwarf-shoot needles and made some distortion where they were feeding (see plate 1); alatoid and apteriod nymphs developed and became adult in about four weeks. The alate sexuparae flew off between mid-May and early June, while the apterous progrediens remained on the same needles and laid their eggs. The progrediens and progeny produced copious masses of wax-wool and honeydew. When a heavy infestation occurred on the foliage the tree was transformed into a blue colour as a result. This was seen to occur at Alice Holt in the period June to early September.

The galls of *A. laricis* have only been found on *Picea sitchensis* during this study. They are unlike those of any other British species in having a creamy colour, a waxy texture and a globular shape, 1 to 1.5 cm diameter (see plate 2). Growth of the shoot often continued beyond the galled portion. During 1965, 1966, 1967, these galls developed rapidly between late May and early June, and opened between the period 13th June to 20th July.

Adelges viridana (Cholodkovsky, 1896)

This species has not been recorded as occurring in Britain before and its presence has only been recognised during the course of this work (Carter, 1969). It is more than possible that it has been in Britain for some time and has been confused with *Adelges viridis* (Ratzeburg) since their first instars, and in particular the adult apterae, are quite similar.

Cholodkovsky (1896) described this larch-feeding species as *Chermes viridanus;* Börner (1908) temporarily placed it in the genus *Pineus* because of similarities in the wax glands of the early instars. More recently Börner and Heinze (1957) have placed this species in the genus *Cholodkovskya* together with *Chermes viridula* (Cholodkovsky) on *Larix russica*, and *Adelges oregonensis* (Annand) on *Larix occidentalis*. Inouye (1953) records *C. viridana* from Japan and has given illustrations of the feeding sites on *Larix*. Heinze's (1962) description of the species is based on the arrangement of the wax gland plates of the first instar sistens; he also describes the unstalked eggs which are laid in fissures on the stem. No forms have been recorded as feeding on *Picea*, therefore it is considered as an anholocyclic species.

MORPHOLOGICAL OBSERVATIONS

First Instar Sistens
They are 0.63 to 0.68 mm long, yellowish green and are without any cuticular pigmentation. The dorsal head shield is divided into two parts by a longitudinal fissure. On it are four groups of wax gland facets of a characteristic type (see fig. 7). Normally the wax glands are composed of three to six facets surrounding a central hair. Rims of the facets are simple and unthickened. Two other cephalic wax glands occur on individual circular plates and lie between the head shield and the ocellani. The thoracic shield is similarly divided along the mid-notal line; there are five groups of wax gland facets on each half. Two marginal thoracic wax glands are on separate plates. The abdomen has the usual six longitudinal rows of wax glands, still of the same type, but each is on its own circular plate. On the last two segments the plates tend to merge laterally.

Adult Sistens
From 1.8 to 2.5 mm long, yellow or yellowish green and do not conceal themselves under the secreted waxy wool, this is deposited laterally. Prepared specimens on microslides have a rounded anterior

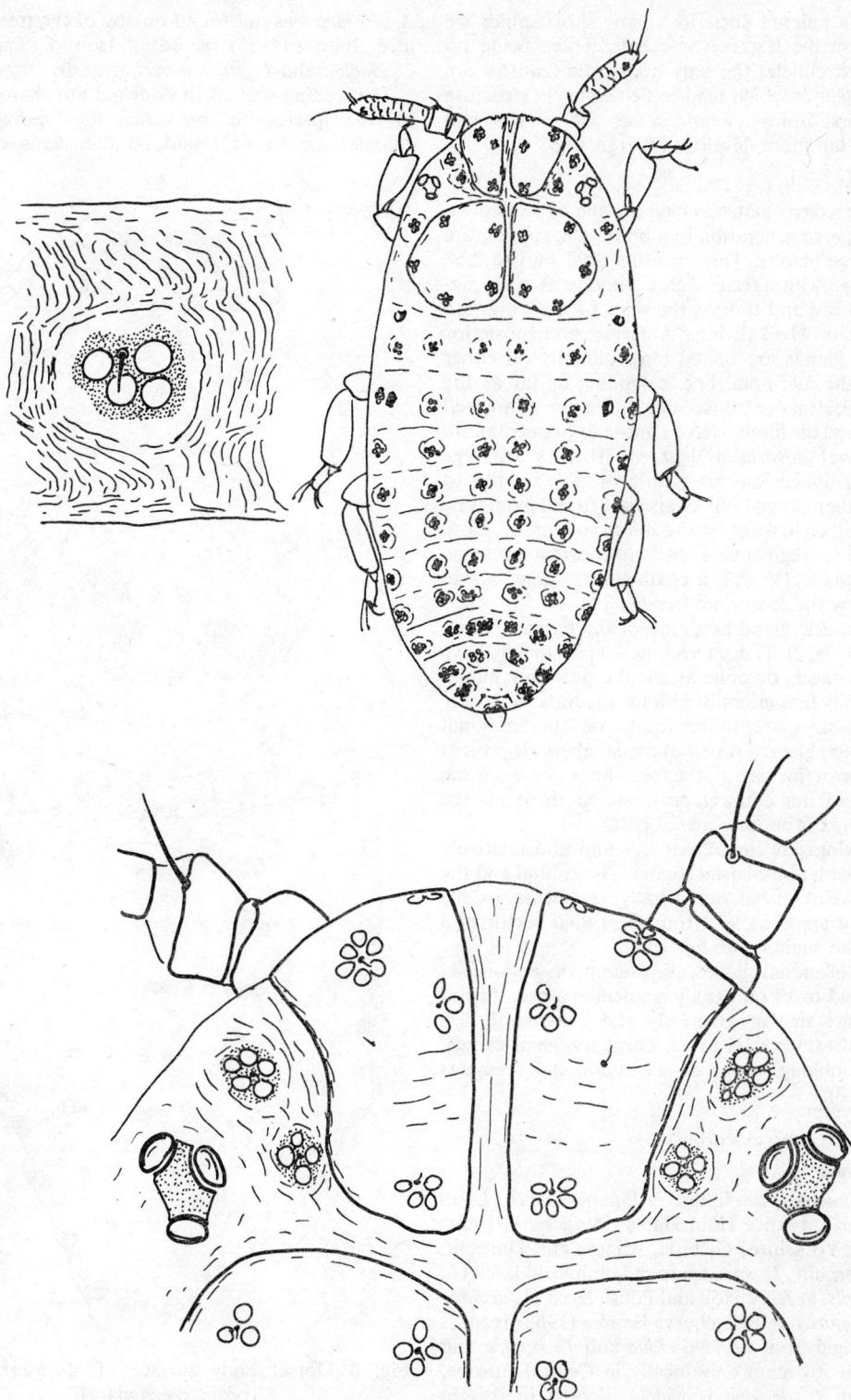

Fig. 7 *Adelges viridana*: dorsal body surface of the first instar sistens (top right); detail of abdominal wax gland facets (top left); detail of facet arrangement of the head (below).

end and a pointed posterior; many short spines are present on the last two segments. There being no pigmented cuticle, the only diagnostic features are the wax glands which tend to be similar in structure to the first instars. Gland areas are very sparse mesially but more developed marginally.

The Alate Forms

The progrediens alatae, which are the equivalent to the sexuparae generation in a holocyclic species, are quite large insects. They measure 1.98 mm to 2.58 mm long, with greenish-grey abdomens and pigmented head and thorax, the veins of the wings are also green when living. Anterior and posterior cephalic glands are united longitudinally on either side of the mid-notal line, extending as far as the median ventral ocellus (see fig. 9). The facets are well defined and the lumina have a granular pigmentation.

Antennal sensoria on segments III to IV are large and their dimensions are a little variable relative to one another. Some individuals were found from time to time which had antennae reduced to four, or, more rarely, three segments. Those with normal antennae have segment IV with a characteristic taper above and below the sensorium (see fig. 10).

Prothoracic gland areas are of the basic arrangement (see fig. 2). The anterior mesial pair mostly have very few facets or none at all; the posterior mesial glands may fuse laterally with the pleurals. A variety of facet sizes are to be found on the marginal prothoracic glands. A pair of mesial glands is present on the posterior part of the mesothorax between the lobes, and immediately posterior to these on the metathorax is another mesial pair.

The wings are long 2.0 to 2.9 mm and relatively narrow with a sub-costal stigma. The cubital and the medial veins curve out distally. A media in the hindwing arises at 90° from the radial sector and curves out slightly (see fig. 9).

Pairs of mesial glands are present on abdominal segments I to VI on weakly pigmented areas. Pleural glands are smaller anteriorly and increase in size between segments III to VI. Large marginal glands, oval or oblong, occur on all abdominal segments (see fig. 8).

BIOLOGICAL OBSERVATIONS

Host Plants

A. viridana has been found on Japanese larch (*Larix kaempferi*) at Alice Holt, Hants; Bedgebury, Kent; Malton, Yorkshire; Glenisla, Angus; and Dunecht, Aberdeenshire. It was also found on hybrid larch (*L. x eurolepis*) at Alice Holt and Polish larch (*L. decidua* var. *polonica*) at Bedgebury. Heinze (1962) records this adelgid from *Larix decidua* and *L. russica* and says that it occurs only locally in Central Europe. Although *L. decidua* is widely planted in Britain

A. viridana was not found on any of the trees examined; Inouye (1953) records it from *L. kaempferi*, *L. gmelini* and *L. gmelini* var. *olgensis*.

The feeding sites of this adelgid are characteristic for the species. In the winter and spring, small colonies are to be found on the stems of trees,

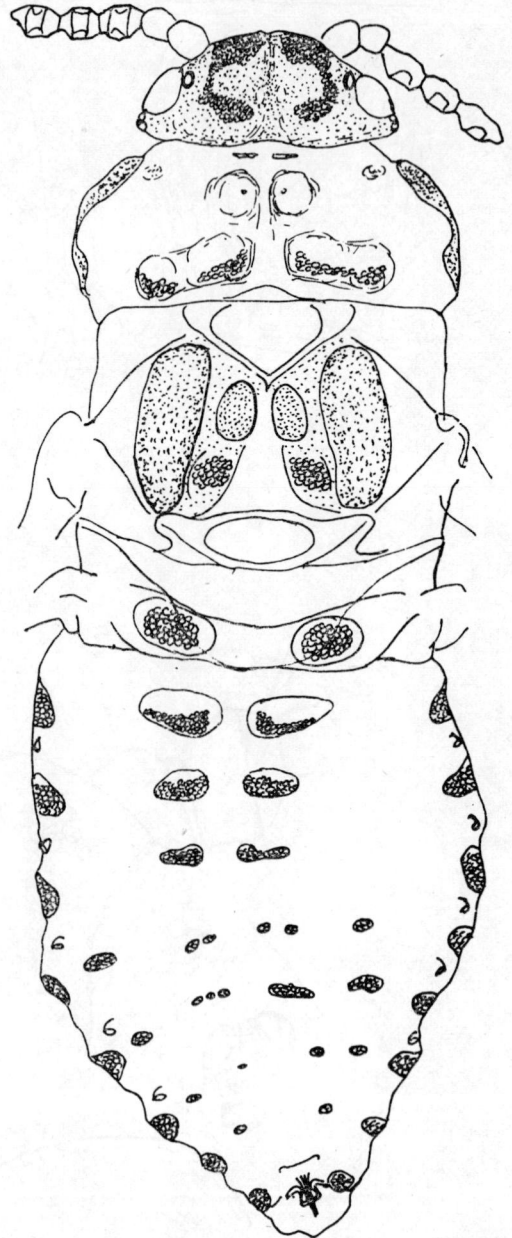

Fig. 8 Dorsal body surface of *Adelges viridana* (progrediens alata).

frequently concealed within bark fissures of old trees or under flakes of bark adjacent to living tissue. On younger trees with smoother bark the colonies sometimes extend up the stem in lines. Colonies do not appear on the foliage until the tender long extension shoots are produced, (Crooke, 1952).

Life Cycle

So far, only parthenogenetic apterous and alate forms have been found on *Larix*. Immature apterae (sistens) occur from September to March; mature apterae have been found with eggs in May and June. The hatching nymphs are very active and crawl to the

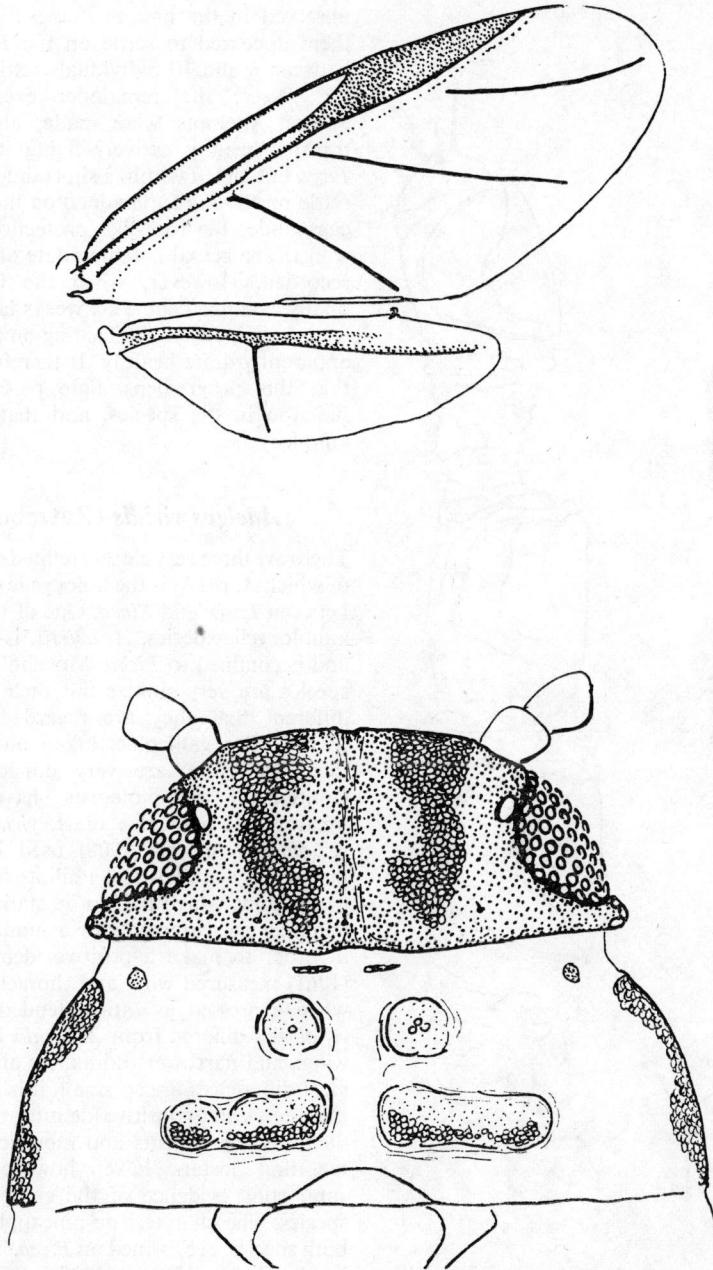

Fig. 9 Wings, head and prothorax of *Adelges viridana*.

tender long shoots to feed. Their growth is quite rapid, and all of those observed developed into alatae which dispersed. Winged adults can be found during June and July. Gaumont (1954) has recorded this species as having three winged generations on *Larix* in the Paris region, thus differing from any other adelgid.

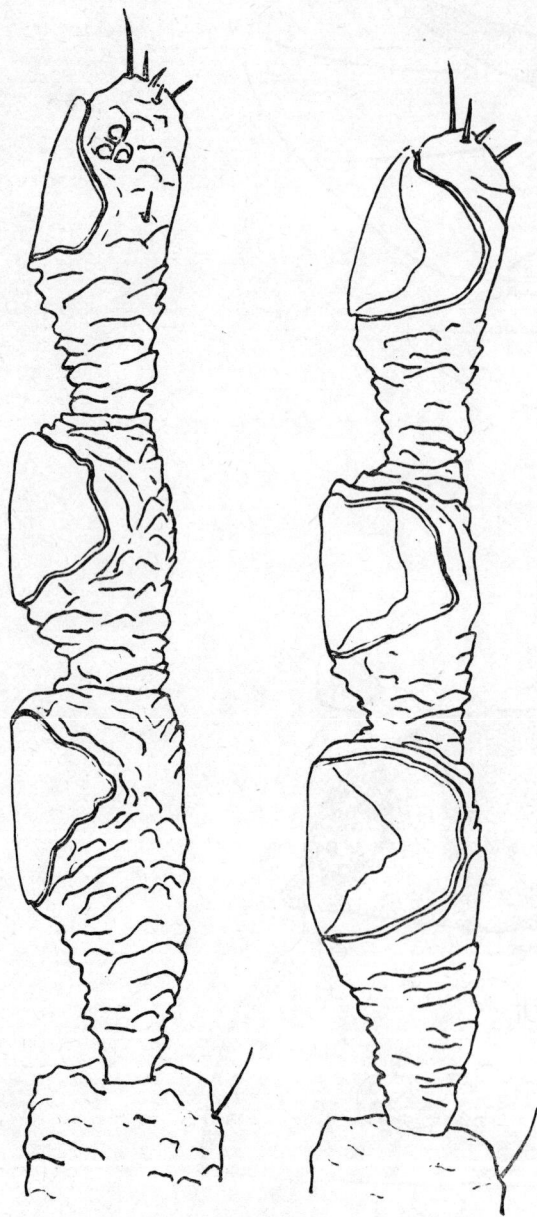

Fig. 10 Terminal antennal segments of *Adelges viridana* (progrediens alata).

Transfer Experiments

About ten colonies with winged nymphs were collected on long shoots and put in a cage on 24th June 1967, together with three uninfested potted plants, one each of *Picea abies*, *P. sitchensis* and *Larix kaempferi*. More than 150 winged adults were produced from the colonies and active flight was observed in the first two weeks of July. None of them appeared to settle on the *Picea* needles, but between 6 and 10 individuals settled and stayed on the *Larix;* the remainder eventually perished. Several attempts were made, also artificially, to transfer healthy actively flying individuals to the *Picea* but all left within a short time. The few that did settle on the *Larix* remained on the needles and laid eggs under the roof-like protection of their folded wings. The actual hatching date of the eggs was not recorded. However, when the Larch was being finally examined some six weeks later a few nymphs were found settled and feeding on the stem and were apparently quite healthy. It therefore appears likely that this progrediens alate performs a dispersal function in the species, and that the life cycle is anholocyclic.

Adelges viridis (Ratzeburg, 1843)

There are three very closely related species of adelgids, of which *A. viridis* is the holocyclic species alternating between *Larix* and *Picea*. One of the closely related anholocyclic species, *A. abietis*, is known in Britain and is confined to *Picea*. Morphologically these two species are very similar but their life cycles are so different that they are regarded as two distinct species. The gallicolae, taken out of context with their life cycle, are very difficult to determine. Continental entomologists have attempted to separate the gallicolae of *A. viridis* from those of *A. abietis*. Börner (1908) used body colour and morphological features; Philiptschenko (1916) used a technique which had to be statistically applied to antennal measurements of a number of individuals in order to make a positive identification; Steffan (1961) measured wing and thoracic features, which, when expressed as ratios, tended to indicate that *A. abietis* differed from *A. viridis* in having broader wings and narrower thorax. As all these morphological differences are so small, it is doubtful if any of them can lead to positive identification of individuals. Steffan's experiments and morphological studies of the first instars have, however, given further supporting evidence of the existence of these two species. The charcteristic pineapple galls, similar in both species, are formed on *Picea*. They are described in detail by Börner (1908), Burdon (1908) and Plumb (1953).

MORPHOLOGICAL OBSERVATIONS
Sexuparae
Length 1.35 to 1.7 mm, width 0.6 to 0.75 mm. The body colour of the living insect is pale green or yellow and the veins of the wings are tinted pale green or yellowish brown.

The head is heavily pigmented and is without any distinct faceted wax gland areas. However, when

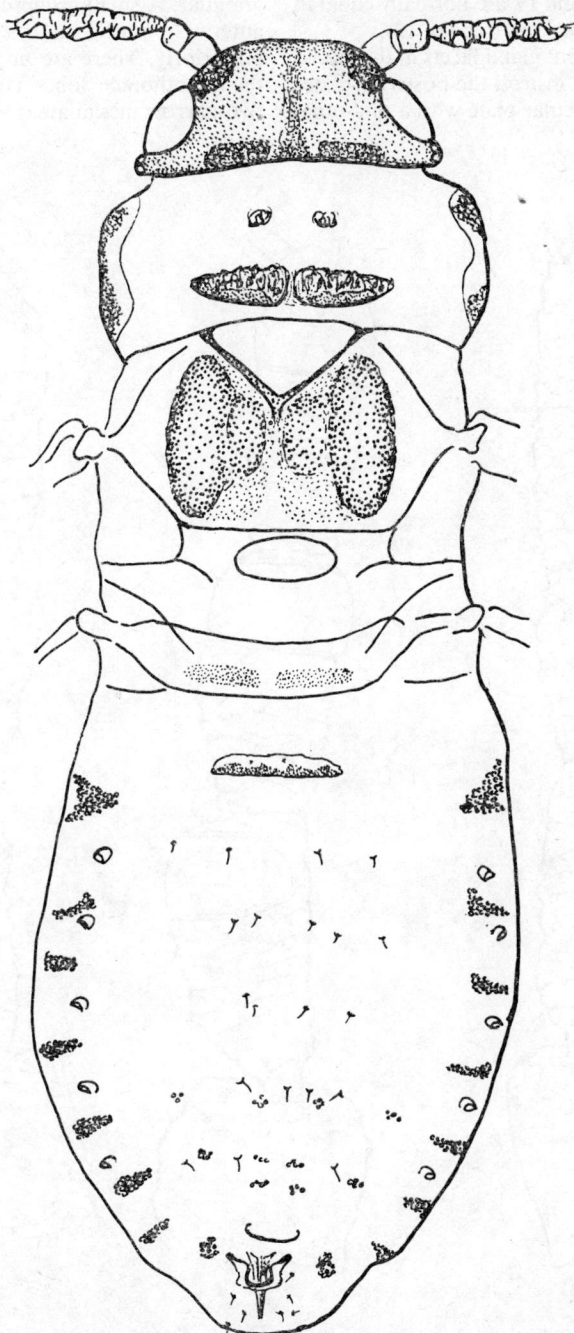

Fig. 11 Dorsal body surface of *Adelges viridis* (sexupara).

examined under phase-contrast illumination, highly glandular areas can be seen near the dorsal posterior margin and near the antennae (see fig. 11). The antennal segments IV and V are not very constricted at their articulating surfaces compared with segment III. The sensoria on III and IV are normally equal in size and have rounded borders (see fig. 12).

The prothorax is without gland facets in the mesial or pleural positions, but instead the posterior plates are fused to form a lenticular plate with a mid-notal furrow (see fig. 11). Fine granulations occur in two patches on the hind margin of the plate, the remainder being covered with irregularly shaped pits. Marginal wax gland areas of the prothorax do have facets, but these are very indistinct and highly granular. A small group of such facets occurs in the anterior position and a larger patch is present posteriorly. There are no wax gland areas between the mesothoracic lobes. On the metathorax there are two narrow mesial areas with fine granules.

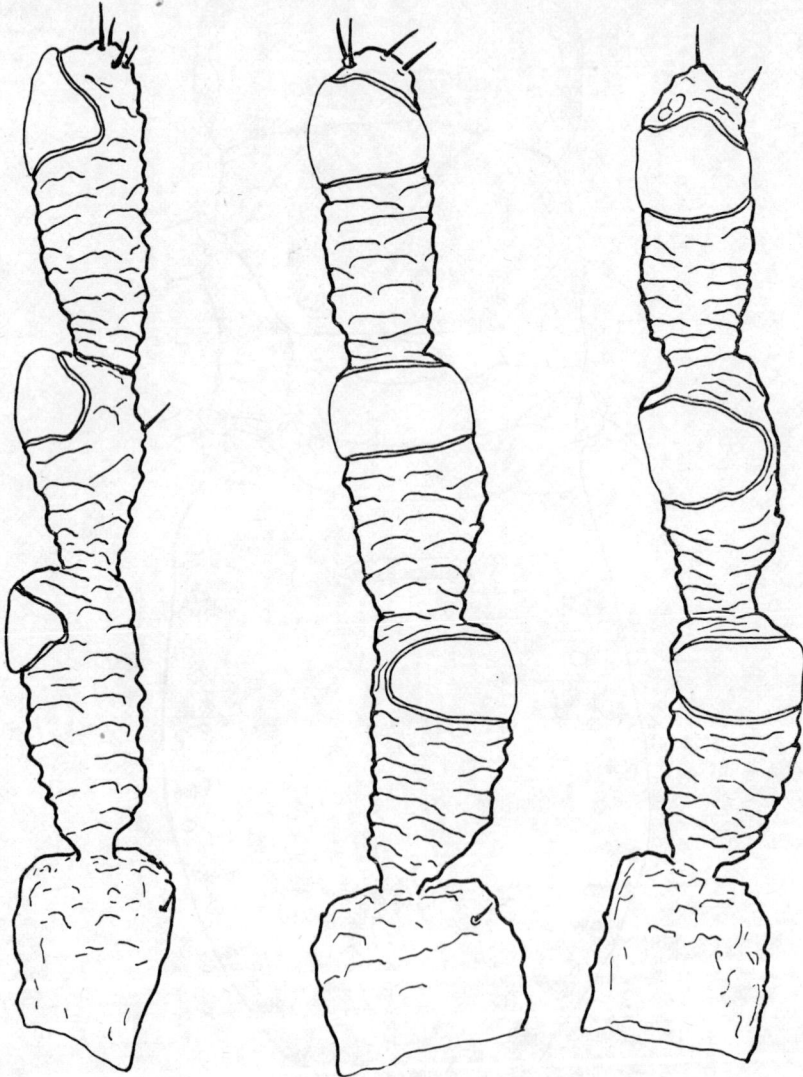

Fig. 12 Terminal antennal segments of *Adelges viridis* (sexupara).

The wings of *A. viridis* and *A. abietis* are very similar (see fig. 13). In both species the forewings are broad, having the costal veins strongly bowed as far as the stigma; the straight hind-wing media are short and arise from the radial sector at about 90°.

In *A. viridis* the forewings measure 1.8 to 2.15 mm long. A line of pigment runs roughly parallel to the hind margin of the sub-costal vein crossing the bases of the other three veins. The cubital vein is sigmoidally curved and the anal vein is strongly arched. The hindwings measure 0.9 – 1.2 mm long.

Both plates and facets on the abdominal mesial areas are lacking, with the exception of the first segment. On it a fused transverse bar occurs that has faint granulations on the posterior half (see fig. 11). In the pleural position on the abdomen, the wax gland facets are few in number and only occur in the posterior half. Prominent marginal wax gland areas occur on segments 1 to 8; there is also a small area sometimes on the 9th segment. The marginal gland

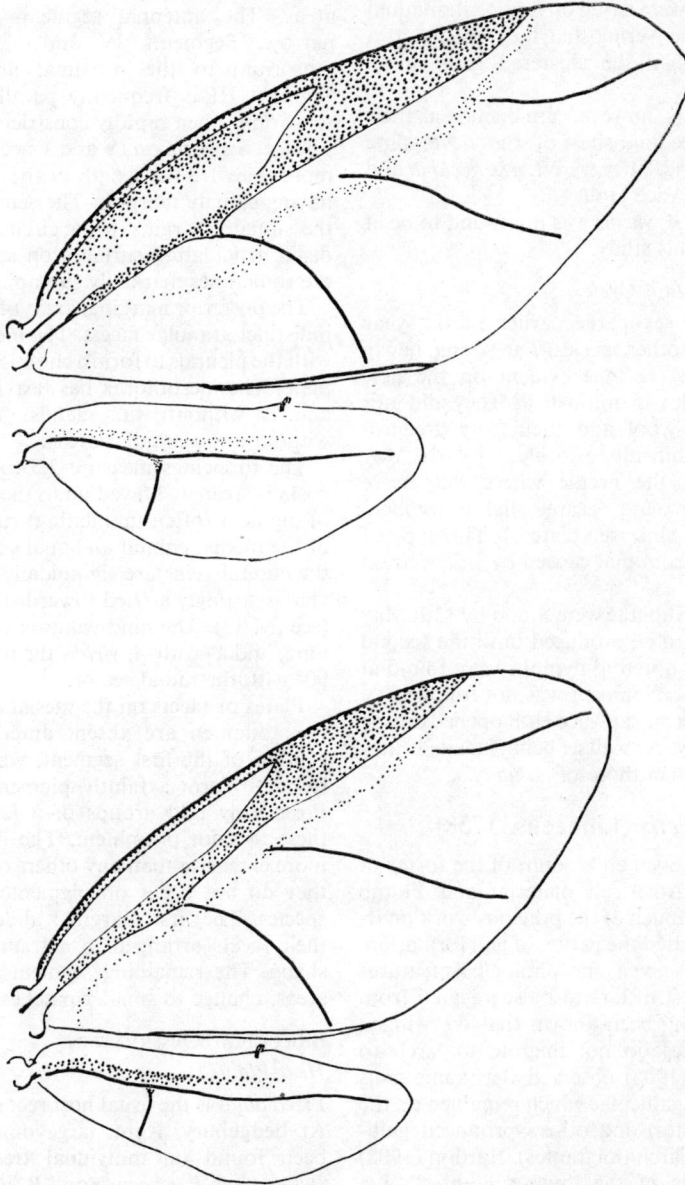

Fig. 13 Wings of *Adelges abietis* gallicola (above) and *A. viridis* sexupara (below).

areas on the first segment are the largest – the facets being grouped together to form a crescented or triangular shape. The remaining marginal glands are frequently crescented at first changing to smaller, narrower patches posteriorly (see fig. 11).

BIOLOGICAL OBSERVATIONS

Host Plants

Picea abies and *P. sitchensis* were thought to be the primary hosts at Alice Holt. A few galls were also found on *Picea orientalis* at Bedgebury (see plate 3). These identifications were based on finding individual isolated galls and discovering that they opened at a much earlier date than the clustered galls of *A. abietis*.

The secondary hosts, however, are definite as their needles were the feeding sites of the developing sexuparae in the spring. They were *Larix decidua* and *Larix* x *eurolepis* at Alice Holt.

The occurrence of *A. viridis* was not found to be at all frequent during this study.

Observations on the Life Cycle

A. viridis produced sexuparae earlier in the year compared with most other species. The young, newly hatched nymphs first became evident on the new growing larch needles in mid-April. They did not produce much wax-wool and their pale greenish colour made them difficult to locate. In early May the actual spot on the needle where they were feeding turned yellow and became slightly swollen, producing a double kink (see plate 4). This type of kinking is different from that caused by *A. laricis* on larch needles.

In 1967 the first sexuparae were found by 13th May and they continued to be produced until the second week in June. A few apteroid nymphs were found at this time but their development was not followed.

The galls on *Picea abies* at Alice Holt opened during the first week in July. As well as being isolated, they are possibly larger than those of *A. abietis*.

Adelges abietis (Linnaeus, 1758)

Annand (1928) has given an account of the forms of this adelgid from American material and Plumb (1953) has reviewed much of the previous work on its biology and investigated the nature of gall formation. Steffan (1961) has given morphological features which enable the first instars to be segregated from *A. viridis*. It has long been known that the winged forms of this species do not migrate to larch to reproduce. Börner (1908) observed that some galls on spruce produced gallicolae which remained on the spruce (non-migrantes) and others produced galli-colae which flew to larch (migrantes). Burdon (1908) notes that the galls of the "non-migrantes", i.e. *A. abietis*, opened at a much later date than *A. viridis*.

As no extensive dispersal takes place from the parent tree, large numbers of *A. abietis* galls can occur on individual trees which can cause damage to larger nursery plants and specimen trees. This species is, therefore, anholocyclic, having just two morphs – pseudo-fundatrices and gallicolae.

MORPHOLOGICAL OBSERVATIONS

Gallicolae

Length 1.85 to 2.2 mm, width 0.85 to 0.99 mm. The head is heavily pigmented without any wax gland areas. The antennal segments are elongate and narrow. Segments IV and V taper below the sensorium to the proximal articulating surface. Segment III is frequently parallel-sided below the sensorium, then rapidly constricts to the articulating surface. Sensoria on IV and V are quite large, usually more than 1/3 the length of the segment, and often having sinuate margins. The sensorium on III lacks the sinuate margins, being circular and smaller. The distal articulating surfaces on segments III and IV are sometimes markedly sub-apical (see fig. 14).

The posterior marginal areas of the prothorax have indistinct granular facets. The mesial areas are fused with the pleurals to form a chitinised lenticular shaped plate. The mesothorax has heavily pigmented lobes and is without wax glands. The metathorax is without any glandular areas.

The forewings measure 2.5 to 2.8 mm long. The costa is strongly bowed up to the stigma. A faint line of pigment (often undulating) runs across the bases of the media, cubital and anal veins. The media and the cubital veins are sigmoidally curved but the anal vein is strongly arched towards the posterior margin (see fig. 13). The hind wing measures 1.3 to 1.6 mm long, and as with *A. viridis* the media arises at about 90° with the radial sector.

Plates or facets on the mesial and pleural areas of the abdomen are absent anteriorly, with the exception of the first segment, where the mesials are fused to form a faintly pigmented transverse bar. Posteriorly odd groups of a few facets occur but these are not prominent. The marginal glands are more extensive than any others on the abdomen, but they do not occur on pigmented areas as in some species. The first marginal abdominal glands have their facets arranged in a triangular or crescented shape. The remaining segments, at first crescented areas, change to smaller patches posteriorly.

BIOLOGICAL OBSERVATIONS

Host Plants

Picea abies is the usual host recorded for this species. At Bedgebury, Kent, large numbers of galls have been found on individual trees of *P. sitchensis P. koyamai*, *P. jezoensis* and *P. glauca* var. *albertiana*. The gall size is variable and probably depends upon

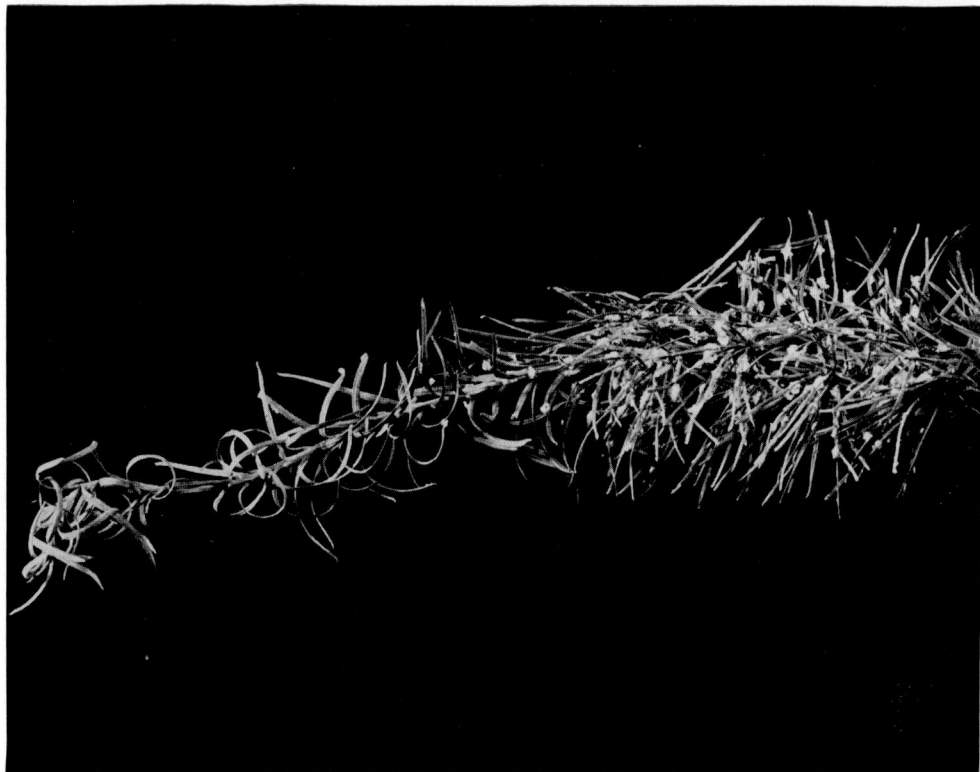

1. Damage to the needles of *Larix decidua* caused by the progrediens of *Adelges laricis*.

2. Unopened gall of *Adelges laricis* on *Picea sitchensis*.

C

3. *Adelges viridis* galls formed on *Picea orientalis.*

4. Kinked needles of *Larix decidua* caused by the feeding activity of the sexuparae and progrediens of *Adelges viridis.*

5. Gall of *Adelges abietis* formed on a vigorous shoot of *Picea sitchensis*.

6. Galls of *Adelges abietis* formed on the shoots of the lower branches of a mature *Picea abies*.

7. Characteristic elongate gall of *Adelges cooleyi* on a shoot of *Picea sitchensis*.

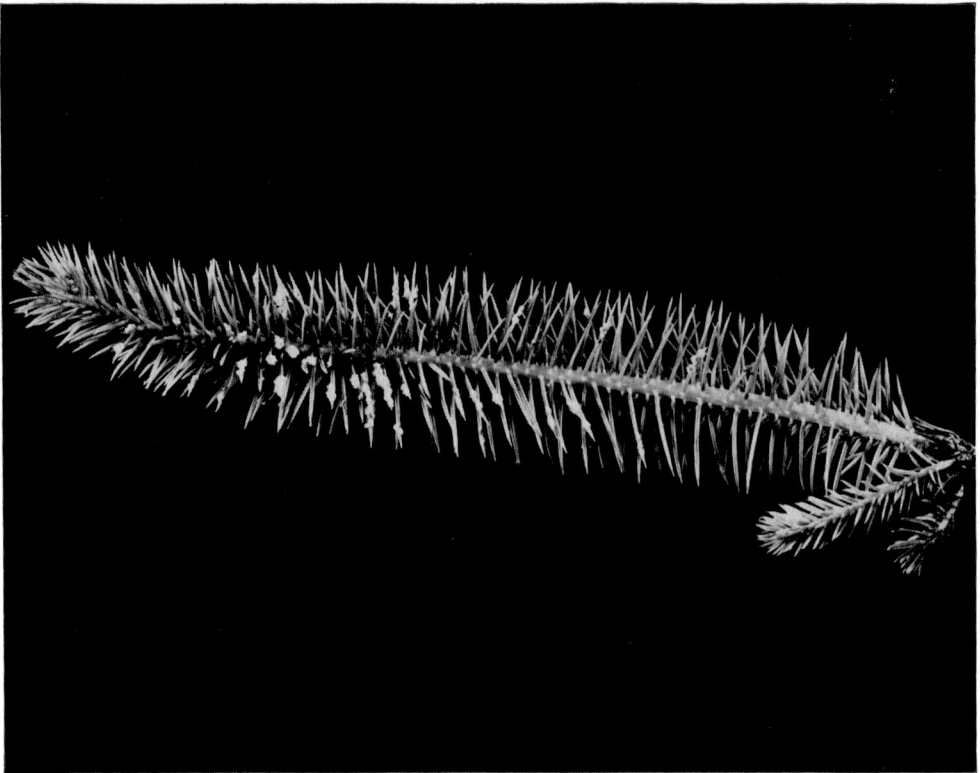

8. Migrant sexuparae of *Adelges cooleyi* settled on the new needles of *Picea sitchensis*.

9. Mature sistens and eggs of *Adelges nordmannianae* on a shoot axis of *Abies cilicica*.

10. Needle distortion and reduced shoot growth of *Abies cilicica* caused by the feeding of *Adelges nord-mannianae*.

1

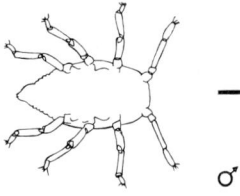

♂

2

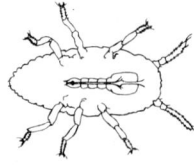

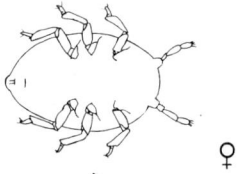

♀

5a

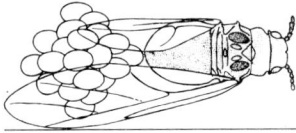

5b

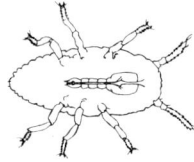

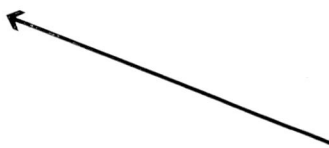

5a

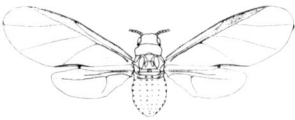

5a

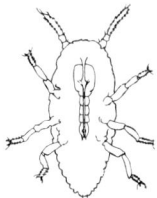

4

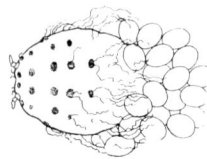

12. Nymph of *Adelges nordmannianae* sexupara on needle of *Abies cilicica*. The large droplets are honeydew excretion.

13. Unopened gall of *Adelges nordmannianae* on *Picea orientalis*.

2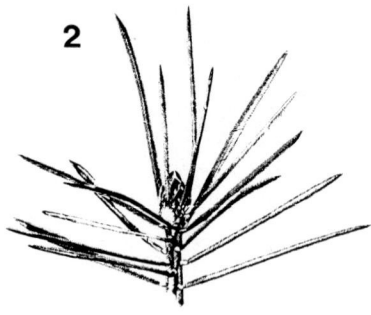

2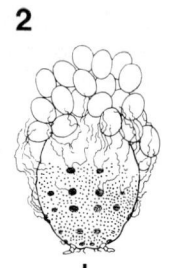

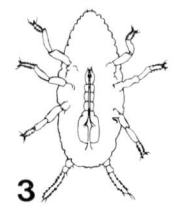

3

3

5b

3

5b

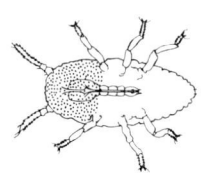  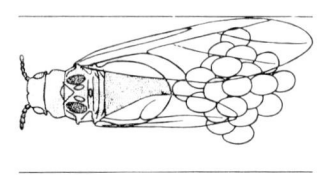

4

3

Plate 11. Life cycle diagram of *Adelges laricis,* a conifer woolly aphid that has five distinct morphs. All the morphs are parthenogenetic, oviparous, females apart from the sexuales which are gamic males and females. Two distinct host plants are involved; the minimum time required to complete such a cycle (outer sequence) is two years. The progrediens apterae give rise to generations of their own kind during the summer. For further details see page 3.

1	sexuales (summer)	**2**	young fundatrix on spruce shoot (autumn)	**2**	developing fundatrix near spruce bud (early spring)	**2**	adult fundatrix (spring)
						3	young gallicola (spring)
5a	sexupara with eggs laid on new spruce needle	**5b**	young progrediens aptera on larch needle (spring and summer)	**5b**	progrediens colonies on dwarf shoots (spring and summer)		
						3	open spruce gall (summer)
5a	sexuparae or progrediens alata fly to spruce (early summer)			**5b**	adult progrediens aptera on larch needles	**3**	gallicolae fly to larch (summer)
5a	young progrediens alata on larch needle	**4**	adult sistens on larch twigs (spring)	**4**	young sistens on long shoots and twigs of larch (autumn and winter)	**3**	gallicola with eggs laid on larch needle

(Plate 11: Life cycle diagram of a conifer woolly aphid.)

14. Gouty shoots on *Abies* caused by *Adelges piceae*.

15. Migrated sexuparae of *Pineus orientalis* settled and feeding on the new shoot tips of *Picea orientalis*.

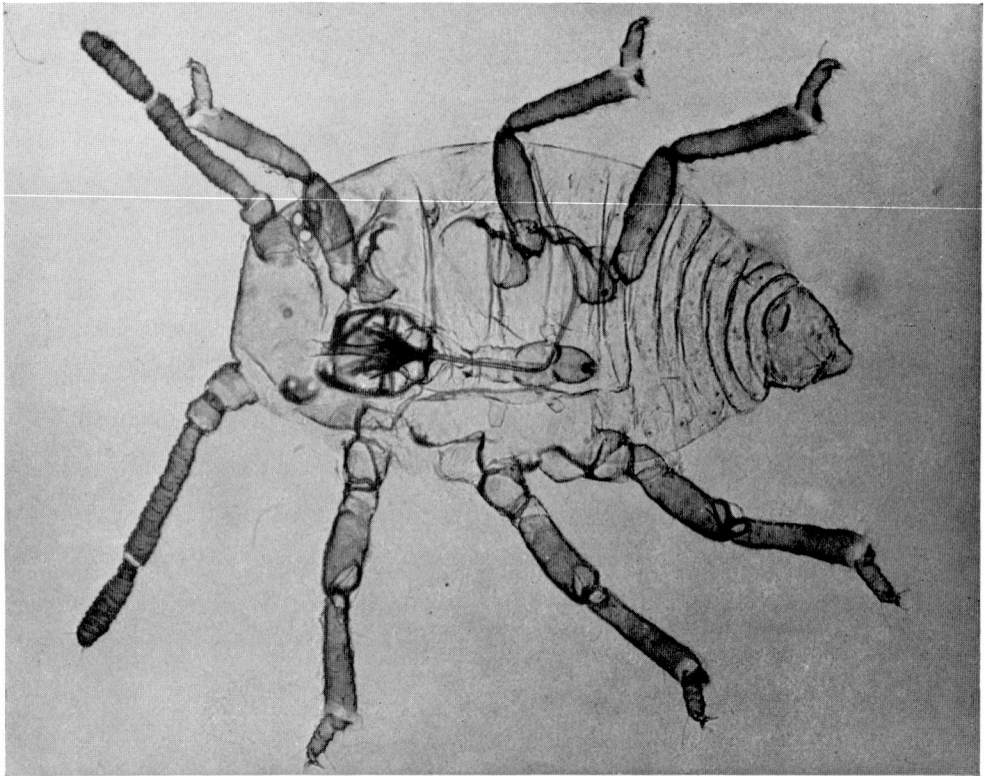

16. Adult male (sexualis) of *Pineus orientalis*. Photomicrograph of a prepared specimen.

17. Adult female (sexualis) of *Pineus orientalis*. Photomicrograph of a prepared specimen.

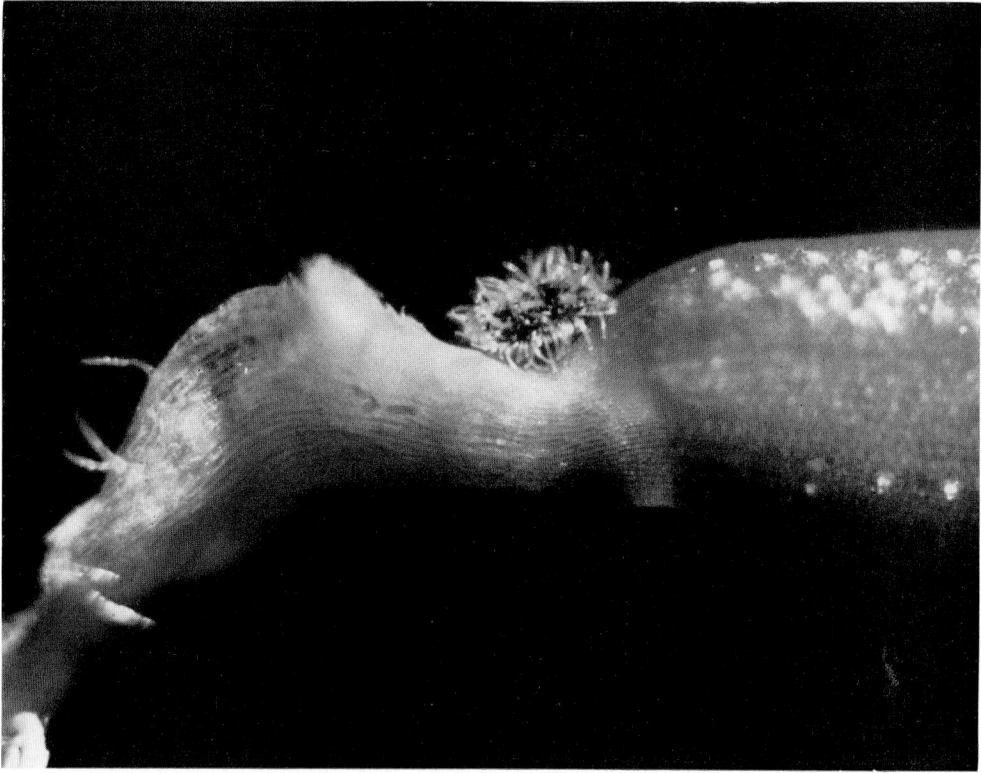

18. First instar fundatrix of *Pineus orientalis* feeding in the axis of the needle base of *Picea orientalis*. (Detached from twig.)

19. Three new shoots influenced by the feeding of *Pineus orientalis* fundatrix (white blob) and progeny on *Picea orientalis*.

20. Three unopened galls of *Pineus orientalis* on *Picea orientalis*.

21. Colony of *Pineus orientalis* on *Pinus sylvestris* as it appeared on 10 May 1966 as a result of an experimental transfer of winged gallicolae the previous summer.

the vigour of the shoots. Those occurring on the lower branches of trees were quite small (see plates 5 & 6).

In 1965, the earliest records for the opening of galls was 2nd and 3rd August. The main time of opening was between 16th August and 1st September at Bedgebury on the species listed above.

During 1966, when the latter part of the summer was particularly dull and wet, the galls on *P. abies* at Alton Forest, Hampshire, had not opened even by the end of September.

A further abundance of galls occurred in 1967 on the same individual *P. koyamai* trees at Bedgebury which were infested in 1965. These galls started to

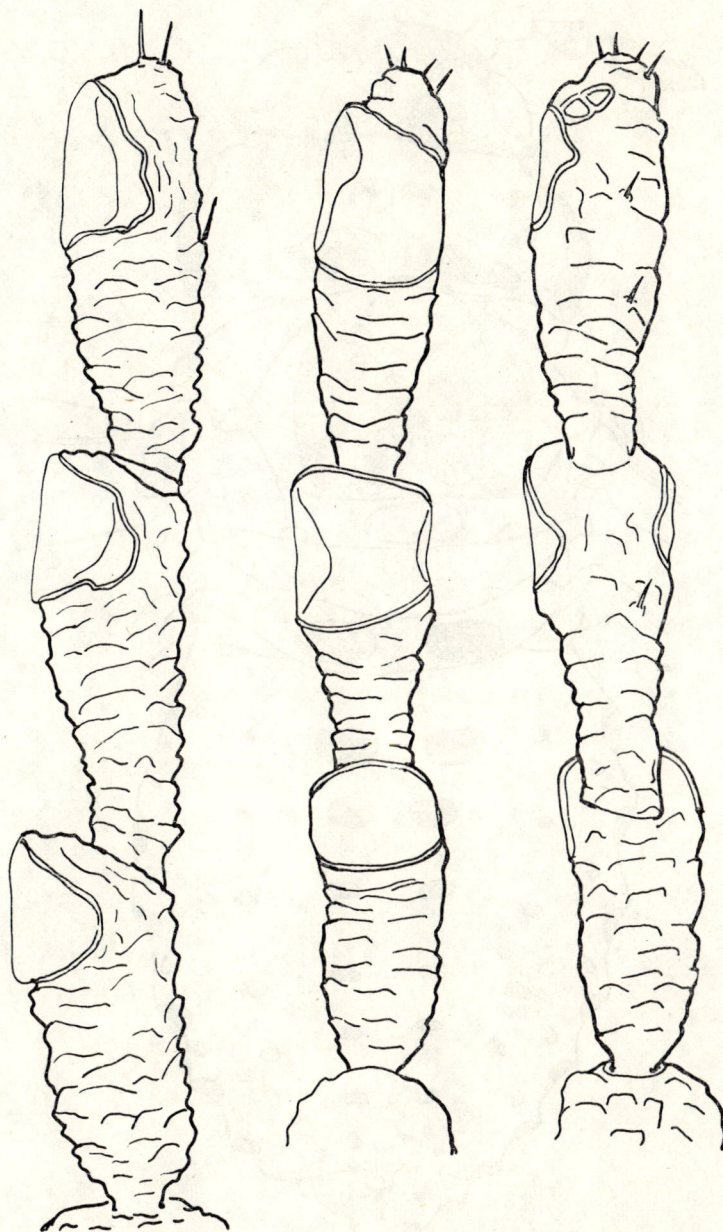

Fig. 14 Terminal antennal segments of *Adelges abietis* (gallicola).

open on the 29th August but many others were still unopened on the 6th September.

Adelges cooleyi (Gillette, 1907)

Descriptions of all the morphs of this species have been given by Chrystal (1924) and Annand (1928) mainly from North American material. At the time of Chrystal's (1924) article, the forms on the *Picea* host were not known in Britain, but since then Cameron (1936) has established that the holocycle

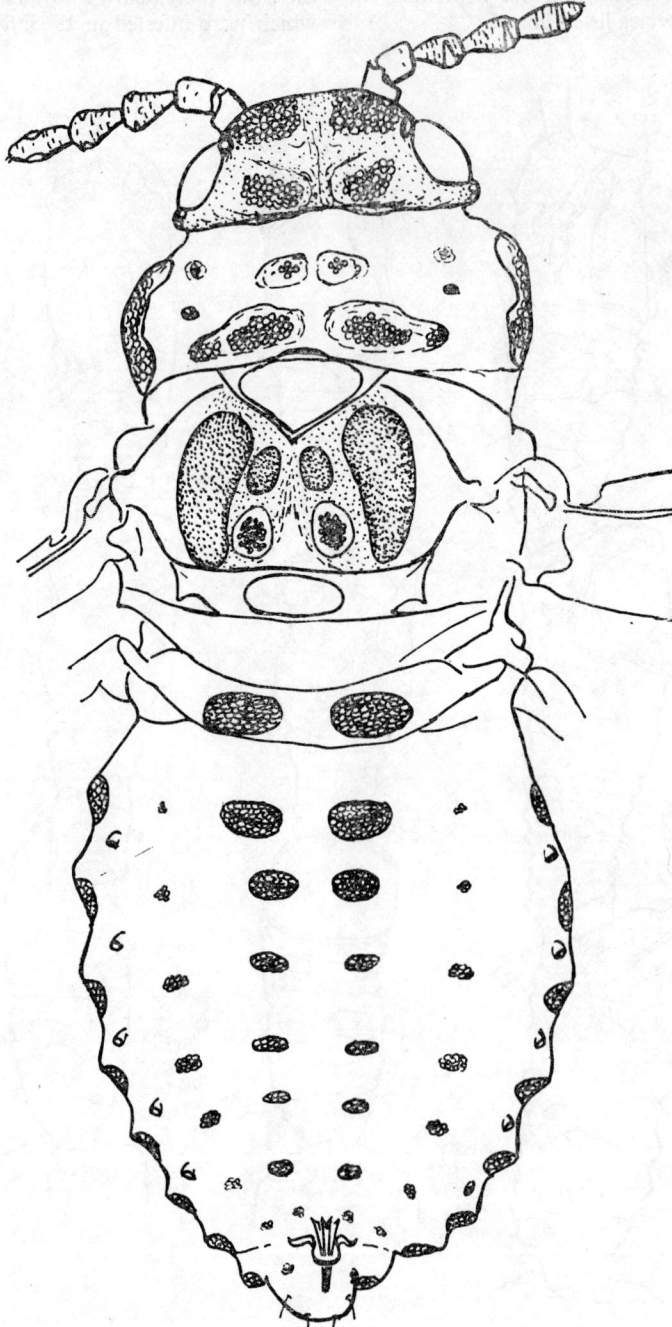

Fig. 15 Dorsal body surface of *Adelges cooleyi* (sexupara).

does occur here and that migration takes place between two commonly grown forest tree species, *Picea sitchensis* and *Pseudotsuga menziesi* in Scotland. Crooke (1960) has investigated the importance of this species in British forestry and given information on damage caused.

MORPHOLOGICAL OBSERVATIONS

Further descriptions are given below enabling the separation of the two winged forms.

Sexuparae
Length 1.2 to 1.7 mm, width 0.55 to 0.66 mm, reddish-brown abdomen. The anterior and posterior cephalic wax glands have large distinct facets (see fig. 15). The antennal segments III to V each have a narrow slit like sensorium which extends only halfway round the segment. The width of segments III and IV is about 2/3 their length (see fig. 16).

The facets on the anterior mesial plates of the prothorax are few, but those on the posterior mesials and pleurals are numerous and frequently run together (see fig. 15). A pair of pale mesial wax glands are very evident between chitinised mesothoracic lobes. Posterior to these are two very large oval areas on the metathorax.

The forewings measure 0.44 to 0.55 mm long. The medial and cubital veins are almost straight, but the anal vein curves towards the thorax (see fig. 18). The hind-wings are 0.26 to 0.33 mm long, with a fairly distinct, straight medial vein which leaves the radial sector at approximately 70°.

Large oval mesial plates occur on the two anterior abdominal segments, these decreasing in size posteriorly. The facets on the plates have polygonal shapes with granular lumina (see fig. 17). The spaces between the facets are, on average, less than half their

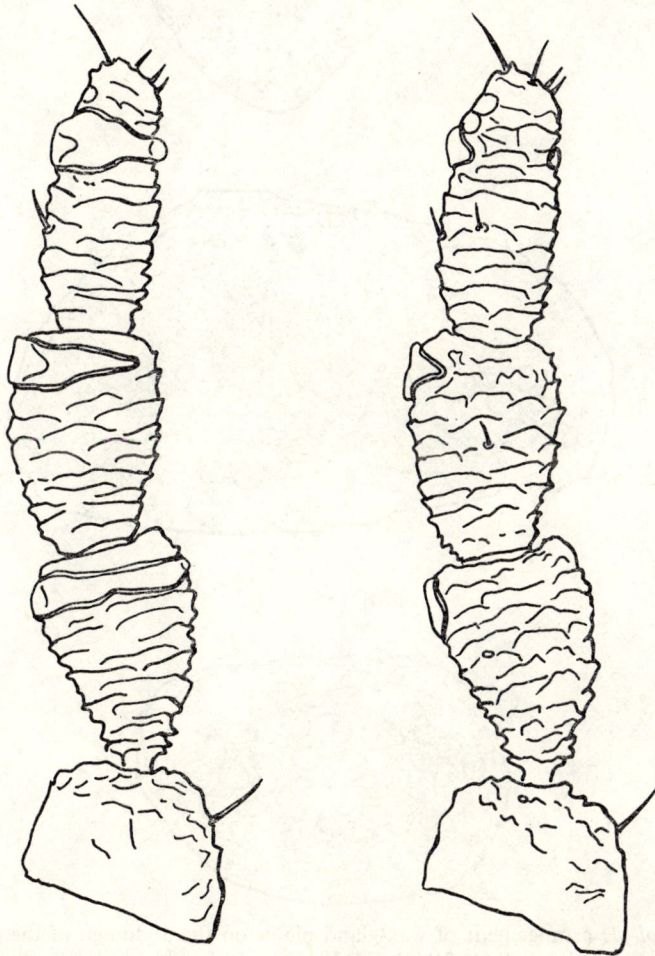

Fig. 16 Terminal antennal segments of *Adelges cooleyi* (sexupara).

diameter. Similar large oval plates are present in the marginal position of segments one to eight. The pleural plates are normally represented by at least a few facets on segments one to eight (see fig. 15).

Gallicolae
These are generally larger than the sexuparae, being 1.7 to 2.5 mm long and 0.9 to 1.25 mm wide, reddish-brown abdomen and wing veins.

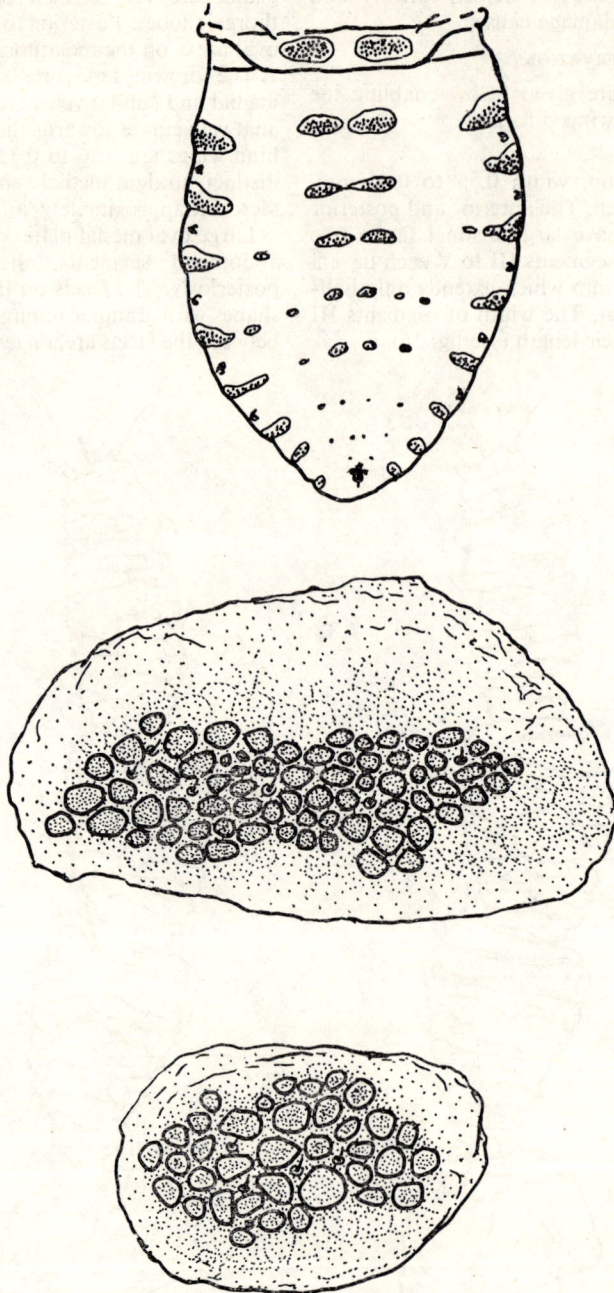

Fig. 17 *Adelges cooleyi*: arrangement of wax-gland plates on the abdomen of the gallicola (top); detail of an anterior mesial wax-gland plate of the gallicola (centre); detail of a plate from the same region in the sexupara (below).

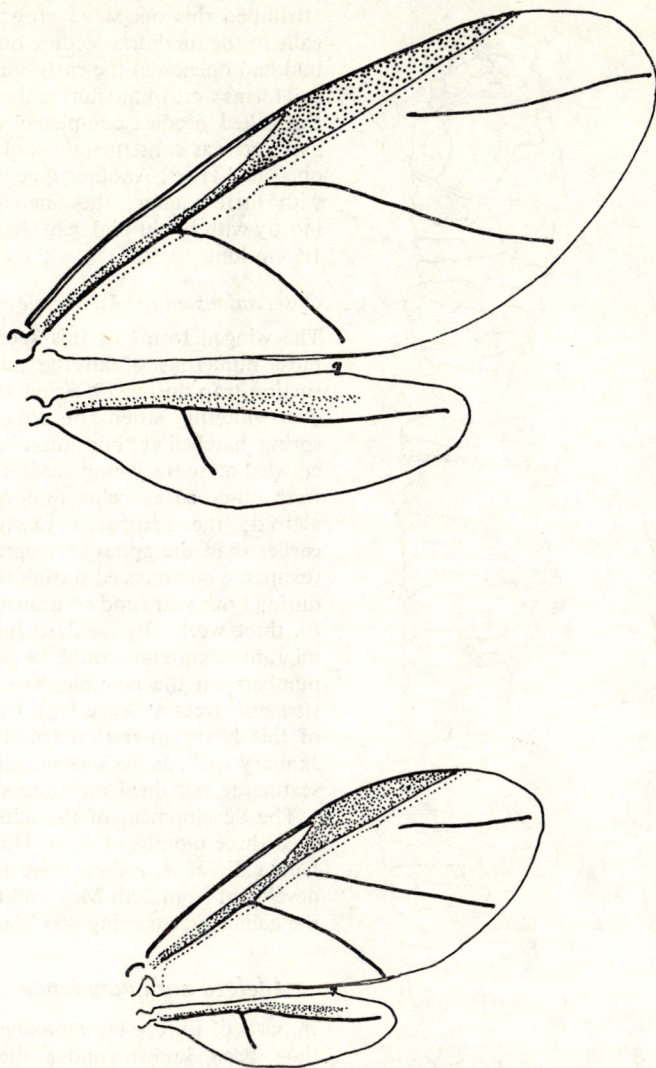

Fig. 18 Wings of *Adelges cooleyi*, gallicola (above) and sexupara (below).

The antennal segments III to V are longer than those of the sexuparae; segments III and IV are carrot shaped, having slit-like sensoria. The sensoria extend slightly more than half way round but do not occupy more than $\frac{1}{4}$ the length of the segment. The widest part of the III and IV segments is less than 2/3 of their length (see fig. 19).

The arrangement of the head and prothoracic wax glands is the same as on the sexuparae. The forewings measure 0.69 – 0.9 mm long; the cubital veins are usually curved sigmoidally (see fig. 18). The hindwing lengths are 0.49 – 0.55 mm, having a distinct straight medial vein making an angle of approximately 60° with the radial sector.

The abdominal wax gland plates differ from the sexuparae by being an elongated shape in the marginal position and tend to fuse in the mesial position. The anterior mesials are quite large but the posterior mesials are frequently lacking; the pleurals are also reduced or even missing (see fig. 17). Granules are clearly visible in the polygonal facet lumina. The spaces between the facets are, on average, less than half their diameter (see fig. 17).

BIOLOGICAL OBSERVATIONS
Galls

Galls on *Picea sitchensis*, when fully developed, have a characteristic elongate form unlike any other

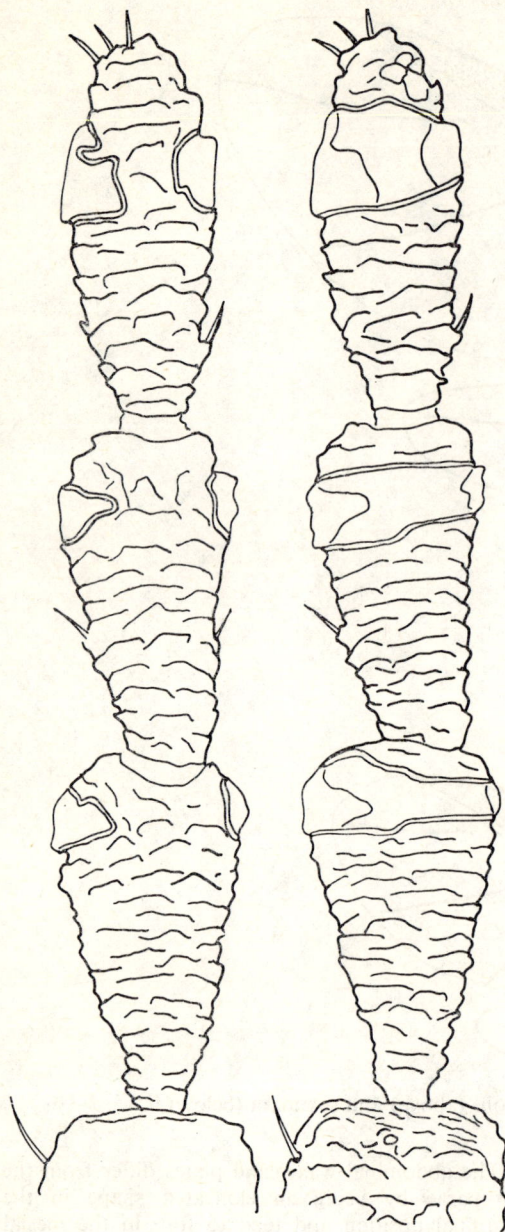

Fig. 19 Terminal antennal segments of *Adelges cooleyi* (gallicola).

adelgid galls in Britain (see plate 7). The gall chambers vary from green to a purplish red while they are occupied by insects.

Their associated needles are not reduced in length as in many other species. In most cases examined the needles had swollen (galled) bases only on one side of the shoot, thus causing the shoot to twist or, in the case of very long galls, to grow spirally. Plumb (1953)

attributed this one-sided growth in *Adelges abietis* galls to the fundatrix feeding on that side before the bud had opened in the early spring. Several atypical gall forms were found during this study. One type had the galled needles completely encircling the shoot axis and was consequently shorter than most of the one-sided types. Another type was found on shoots with little vigour; they had quite short needles. mostly with bright pink gall chambers and were only 1.5 cm long.

Observations on the Life Cycle

The winged forms of this species were by far the most numerous of all the adelgids taken in the suction trap during 1966 and 1967. The eggs of the over-wintering sistens on *Pseudotsuga*, laid in the spring, hatched at "bud-burst" and the nymphs then crawled on to the young needles to feed. The nymphs were observed as being dimorphic – apteroid and alatoid; the sexuparae reached maturity much earlier than the apteroid progrediens. Flight of the sexuparae commenced during the first week in June during both years and continued on favourable days for three weeks. By the 23rd June, 1967, the current migrant sexuparae could be seen settled in great numbers on the new shoots of many small *Picea sitchensis* trees at Alice Holt (see plate 8). Evidence of this heavy migration could still be detected in January 1968, as the wax secretions and parts of the sexuparae remained on more sheltered shoots.

The development of the adult gallicolae took at least three months at Alice Holt in 1966 and 1967. New galls of *A. cooleyi* were found to be partially developed from 20th May and the earliest record of the gallicolae emerging was 21st August.

Adelges nordmannianae (Eckstein, 1890)

In parts of Europe *A. nordmannianae* has for a long time been known under the name of *Adelges nüsslini* (Börner). There has been much confusion between this species and *A. piceae* (Ratz.) due to their both occurring on *Abies* spp. and to their morphological similarities. *A. nordmannianae* has become a serious pest of young European silver firs, whereas *A. piceae*, although European in origin, causes more damage to the North American species. Marchal (1911b) and (1913) clearly established *A. nordmannianae* as a holocyclic species and *A. piceae* as an anholocyclic one. Steven (1917) and Varty (1956) have recorded galls of *A. nordmannianae* on *Picea orientalis* in Scotland but none were found by Chrystal (1925) in the south of England.

Morphological studies of *A. nordmannianae* and related species have been mainly confined to the first instar sistens. Pschorn-Walcher and Zwölfer (1958) and Eichhorn (1958) have described methods

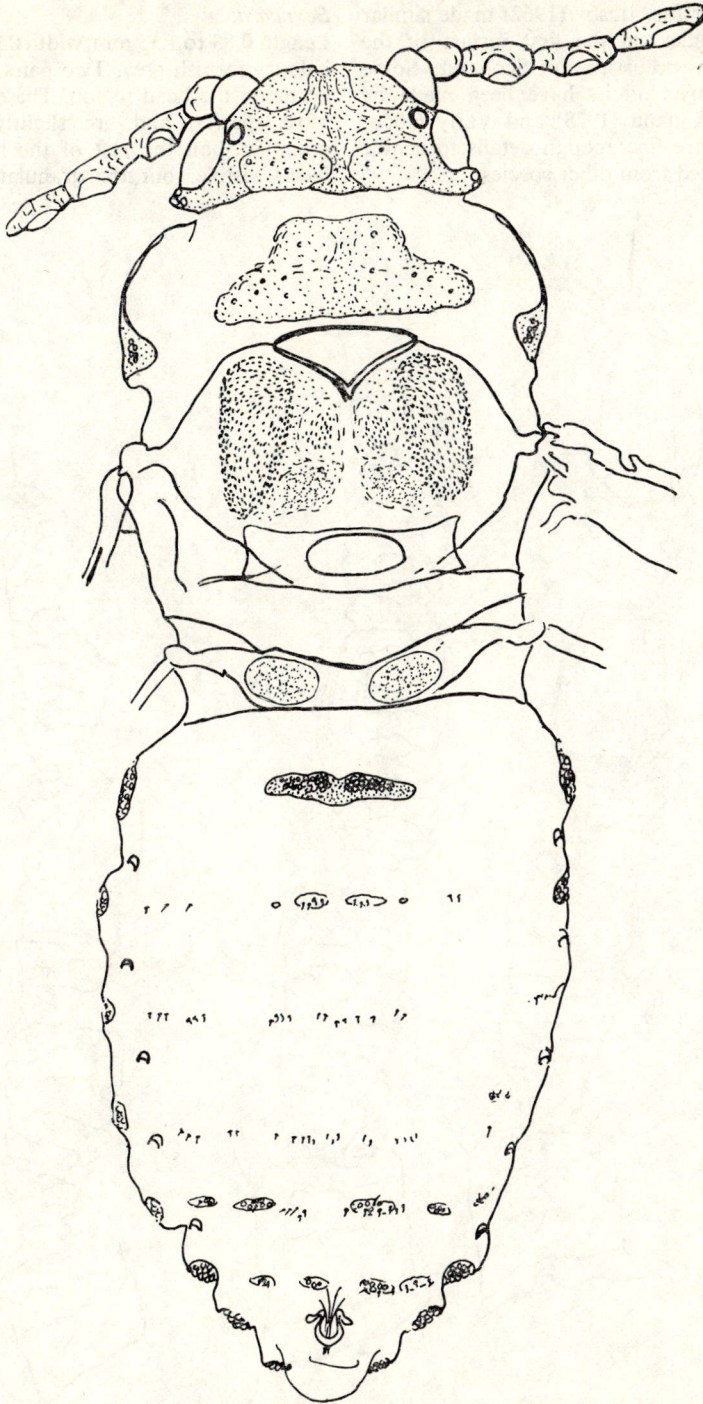

Fig. 20 Dorsal body surface of *Adelges nordmannianae* (gallicola).

of separating the first instars of the European species occurring on Silver firs. Busby (1962) made similar morphological studies on the first instars of the Silver fir adelgids occurring in Scotland. Some details of the winged adults have been made by Marchal (1913), Annand (1928) and Varty (1956) but they do not give fine enough details to enable these to be separated from other species.

MORPHOLOGICAL OBSERVATIONS
Sexuparae

Length 0.83 to 1.15 mm, width 0.38 to 0.44 mm, body colour greenish grey. Two pairs of wax gland areas occur on the head region. These are longer than in most species, and are slightly paler and more granular than the rest of the head. There are no facets visible, but the granulations appear to be

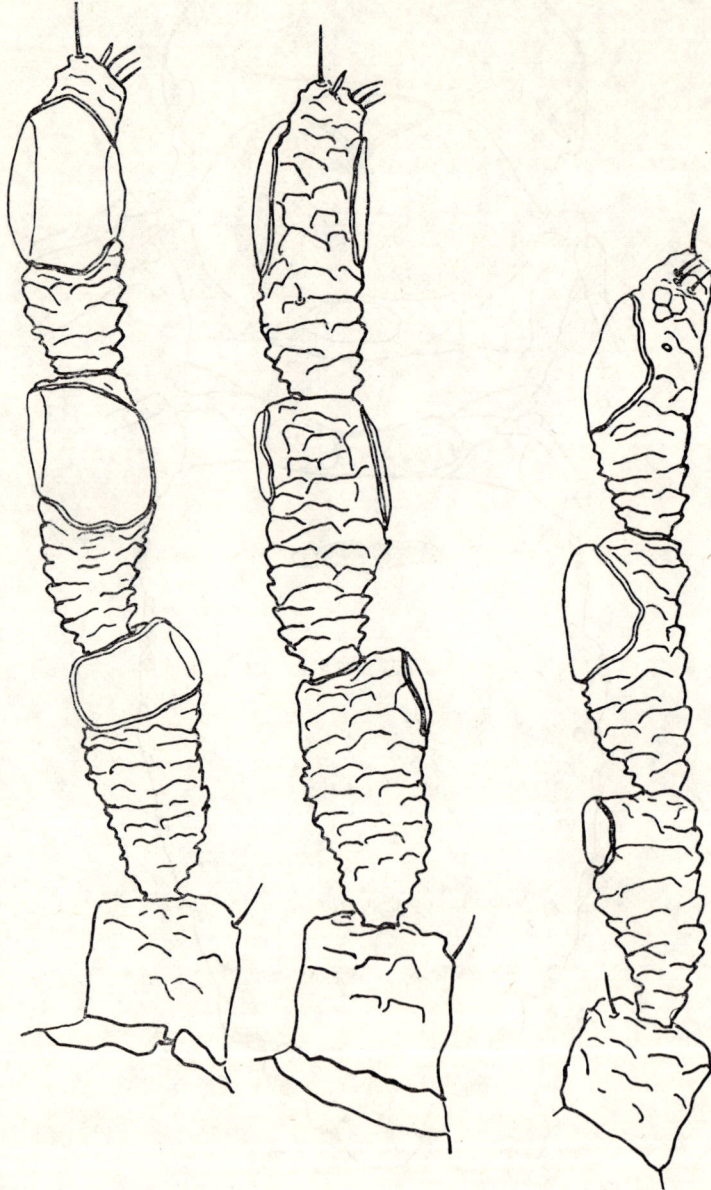

Fig. 21 Terminal antennal segments of *Adelges nordmannianae* (sexupara).

arranged into polygonal areas when seen under phase-contrast illumination (see fig. 20). The antennae have well developed sensoria. Those on segments IV and V extend more than half-way round, and occupy almost half the length of, their segment. The sensorium on Segment III is smaller, barely extending half-way round, and is only about 1/3 the length of the segment. The dorsal surfaces of antennal segments IV and V have angular wrinkles behind the sensoria (see fig. 21).

There are a few indistinct granular facets on the marginal plates of the prothorax. In the mesial position, all the plates are fused into one including the pleurals. There are no distinct facets on this plate. In the posterior position of the mesothorax, between the mesothoracic lobes, paired mesial gland areas are present. These are without any facets, but the granules are arranged polygonally. The mesial wax glands on the metathorax have a similar appearance (see fig. 20). The forewings measure 0.33 to 0.43 mm long and the hindwings measure 0.17 to 0.22 mm long. The media and cubital veins of the forewing are straight, but the anal vein curves back towards the hind margin. There is no media arising from the radial sector of the hindwing (see fig. 22).

The mesial and pleural wax gland areas of the abdomen do not conform to the basic pattern of fig. 2. The first abdominal segment has a roughly lenticular shaped plate in place of the mesial glands. This is faintly granular, and sometimes some small polygonal facets can be made out. The more normal type of facets occur in clusters on segments 5, 6 and 7, occurring irregularly over the dorsal surface. The marginal glands are usually represented on all the segments; those on segments 1, 2 and the posterior are best developed (see fig. 20).

Gallicolae

Length 1.1 to 1.54 mm, width 0.55 to 0.66 mm, body and wings at first greenish, darkening in a few days.

The wax gland areas on the head are much the same as those of the sexuparae, the antennae, however, are more developed. Much larger sensoria occur on segments IV and V which extend half the length of the segment, or even more in the case of segment IV. Segment V is much narrower than the other segments and distinctly cigar-shaped. The sensorium on III is almost half the length of the segment and extends more than half way round it (see fig. 23). Wrinkling on the dorsal surfaces behind the sensoria is strong and angular, and unlike many other species. (About 30% of the gallicolae examined possessed aberrant antennae, usually taking the form of a reduction in the number of segments. In such cases the terminal segment appeared longer and cigar-shaped, having a sensorium of about two thirds the segment length.)

The plates and granulations of the thorax have the same details as the sexuparae with the exception of the prothoracic plate. This has a granular posterior margin, distinct in only some specimens.

The forewings measure 0.49 to 0.55 mm long and have the media and cubital veins almost straight or have a sigmoid curve. The anal vein curves gently back to the posterior margin. The hind wings measure 0.27 to 0.34 mm long. The media is present in this morph, although faint, and leaves the radial sector roughly at right angles.

The abdominal wax gland arrangement is again similar to that of the sexuparae, but with an increase in the number of facets on the posterior segments.

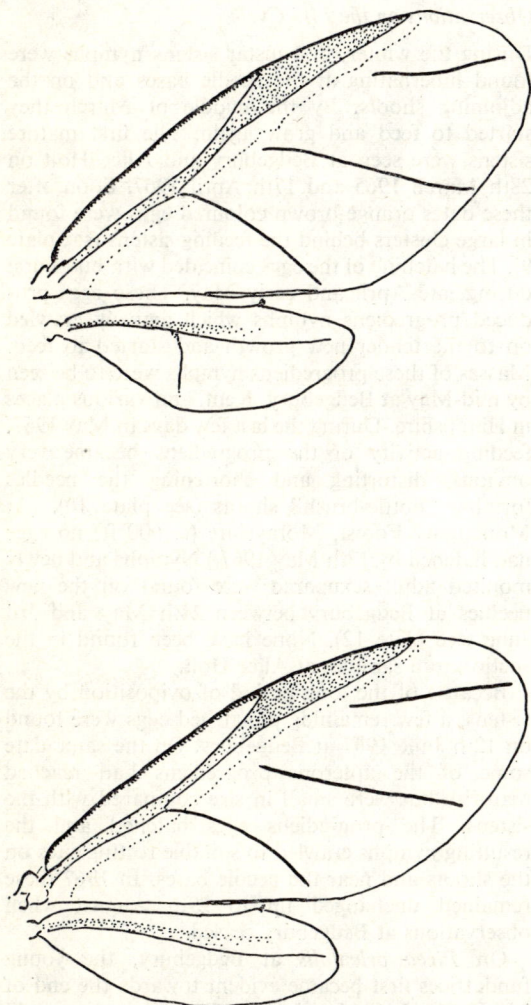

Fig. 22 Wings of *Adelges nordmannianae*, gallicola (above), and sexupara (below).

BIOLOGICAL OBSERVATIONS
Host Plants

This insect was found to be common on the foliage and shoots of *Abies alba*. It was present in great

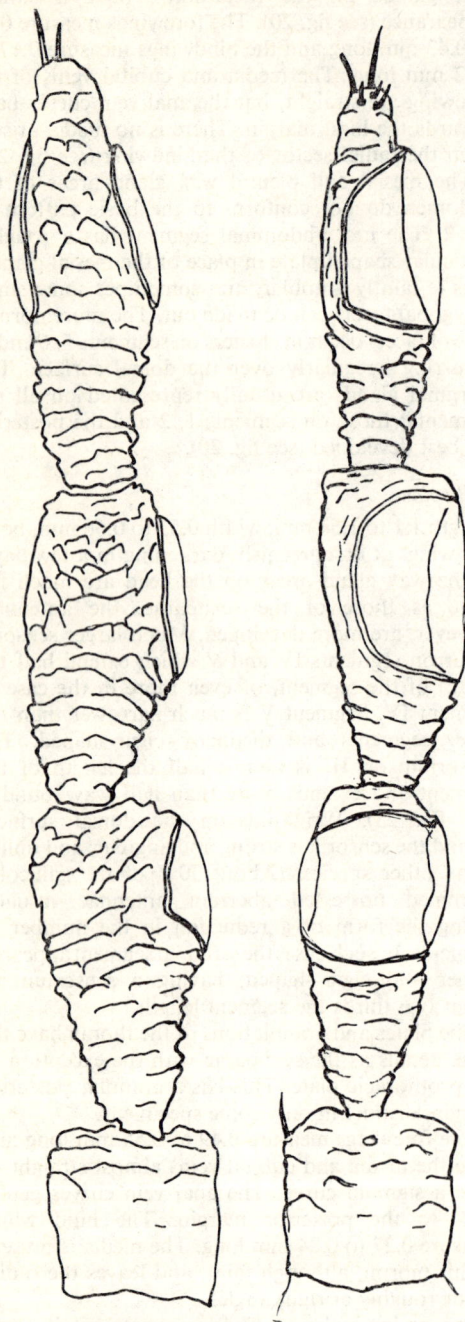

Fig. 23 Terminal antennal segments of *Adelges nordmannianae* (gallicola).

numbers on young trees of *A. nordmanniana* and *A. cilicica* at Bedgebury, causing severe needle distortion (see plate 10). It was also found on the stems of quite old trees at Inverliever, Argyll; Exbury, Hants; and Alice Holt, Hants. These stem colonies do not produce the fluffy wax-wool seen in the colonies of *Adelges piceae*, the individual insects of *A. nordmannianae* having only a peripheral ring of wax wool.

Galls of this adelgid were found infrequently on *Picea orientalis* at Bedgebury, Kent. They were small (1 cm long) and always terminated the growth of the shoot. The needles making up the gall were totally changed to form the gall chambers and were a dull reddish colour before the gallicolae emerged (see plate 13).

Observations on the Life Cycle

During the winter, first instar sistens nymphs were found hibernating at the needle bases and on the adjoining shoots. By the middle of March they started to feed and grow again. The first mature sistens were seen at Bedgebury and Alice Holt on 28th March 1965 and 17th April 1967. Soon after these dates orange-brown coloured eggs were found in large clusters behind the feeding sistens (see plate 9). The hatching of the eggs coincided with bud-burst during late April and early May. These eggs produced progrediens nymphs which actively crawled on to the tender new growth and started to feed. Masses of these progrediens nymphs were to be seen by mid-May at Bedgebury, Kent, and various places in Hampshire. During the last few days in May 1967, feeding activity of the progrediens became very obvious, distorting and shortening the needles forming "bottle-brush" shoots (see plate 10). (At Monaughty Forest, Morayshire (c. 600 ft) no eggs had hatched by 17th May 1967.) Nymphs and newly moulted adult sexuparae were found on the new needles at Bedgebury between 24th May and 3rd June (see plate 12). None have been found in the suction trap samples at Alice Holt.

Because of the long period of oviposition by the sistens, a few remaining unhatched eggs were found on 12th June 1967 at Bedgebury. On the same date some of the apterous progrediens had reached maturity but were small in size compared with the sistens. The progrediens eggs hatched and the resulting nymphs crawled to suitable resting sites on the shoots and near the needle bases. In 1967 these remained unchanged up to 29th August when observations at Bedgebury ceased.

On *Picea orientalis* at Bedgebury, the young fundatrices first became evident towards the end of March (1965, 1966 and 1967). Their presence could be detected by searching the undersides of the buds for white wax-wool secretions but their identity had to be determined by microscopical examination. Towards the end of April and early May discrete blobs (about 1.5 mm diameter) of wax-wool were to be seen, partially concealing the adult fundatrix and eggs. The galls first appeared during the end of May. The emergence of the gallicolae was rather variable – 20th July 1965, 2nd August 1966 and 19th-30th June 1967.

Transfer Experiments

(1) *Abies alba* foliage with young nymphs was collected from Exbury, Hampshire on 10th May 1967. Between 25th and 29th May about thirty sexuparae were produced and were offered a potted *Picea orientalis* plant in a cage. Approximately a third of the sexuparae settled on the needles and laid an average of 5 eggs each. The eggs hatched and sexuales were seen on 26th June, but they later died and no progeny resulted.

(2) Infested shoots of *Abies cilicica* were collected from Bedgebury on 23rd March. At this time some of the overwintering sistens had matured and had started to lay eggs. These eggs were transferred to a potted *Abies alba*. By 29th May the resulting insects were almost mature and it was seen that almost 100% were developing into sexuparae. A similar collection and transfer of eggs was made on 26th April, but this time only 20% developed into sexuparae, the rest became apterous progrediens. The sexuparae from both of these sources were released in a cage containing two year old potted seedlings of *P. orientalis*, *P. abies* and *P. sitchensis*. Here again some of the sexuparae settled on the *P. orientalis* but no fundatrices resulted.

(3) Ten galls were collected from *Picea orientalis* at Bedgebury on 12th June. The first of these started to open on the 19th June in the insectary and the winged gallicolae started flying from the 25th. These were released in a cage containing potted *Abies alba* plants. In spite of there being over 200 gallicolae present, very few readily took to the *Abies*, but a few viable eggs were produced and the resulting sistens nymphs settled on the stems for the remainder of the year.

Discussion

Eckstein (1890) gave an account of two species of adelgids on *Abies*. He did not consider the detailed morphology of the insects but gave biological information. One of these insects he recorded as occurring on the shoots and twigs of *Abies nordmanniana*. This species he named as *Chermes nordmannianae*. The characteristic feeding sites and the damage to the needles described are identical to the observations made during this study on what was hitherto known in Britain as *A. nüsslini* (Börner 1908). Eckstein (1890), in the same article, distin-

guished the other species, *A. piceae* (Ratzeburg 1843), but its feeding sites and also the abundance of wool produced only on the stems and branches. *Dreyfusia nüsslini* of Börner (1908) has long been accepted as a synonym of *Chermes nordmannianae* (Eckstein 1890), and it therefore conforms to the rule of priority in nomenclature to call this insect *Adelges nordmannianae* (Eckstein 1890). This name has already been adopted by continental entomologists but not so far by those in English-speaking countries.

Adelges piceae (Ratzeburg, 1843)

This species, known as the Balsam Woolly Aphid in North America, is considered to be of European origin. It is only known to feed on true firs (*Abies* spp.) and has become a serious threat to many of the forests in Canada. Recent studies by Pschorn-Walcher and Zwölfer (1956 and 1958) and Eichhorn (1957 and 1958) on the ecology and morphology of the Silver fir adelgids in Central Europe have proved that there are three distinct species (*A. piceae*, *A. nordmannianae* and *A. merkeri* Eichhorn). Pschorn-Walcher and Zwölfer (1958) found there were further differences between the specimens of *A. piceae* from North America and Europe. Apart from having habitat differences, the two forms f. *typica* and f. *canadensis* could also be separated on morphological characters to be found in the first instar. Busby (1962), while making a similar investigation in Scotland found two forms, but regarded the stem feeding form on *Abies grandis* as distinct from f. *canadensis* and named it f. *grandis*. However, further material from other parts of Britain examined by Busby (1964) suggests that f. *canadensis* and f. *grandis* are probably synonymous.

Morphological Observations
Alatae

Descriptions of the alatae have been given by Annand (1928). The features of *A. piceae* alatae he gives that separate it from *A. nordmannianae* are: the possession of smaller, rounder sensoria on the antennal segments (particularly III and IV) and in having wrinkles on the dorsal side of antennal segment V that are not strongly arched. The abdominal wax-gland areas of this species are weakly chitinised or even absent altogether in the mesial or pleural areas, their position represented in some cases by just a seta (Varty 1956). Marchal (1913) also figures these features.

Biological Observations
Life Cycle

The production of alate forms in an anholocyclic species such as this is of some interest in relation to its spread into new areas. Marchal (1913) observed such an incidence in France and revealed the alate form from Silver fir as a viable dispersive morph in an anholocyclic life cycle. Alate forms in Britain appear to be of rare occurrence due to the poor synchronisation of egg hatching with the production of new shoots on the trees in the spring. Varty (1956) under experimental conditions, reared alatae by providing new shoots as suitable food for the otherwise abortive first instars of the progrediens generation. No alatae were revealed during this present study.

Host Plants

Although this species is known and probably originated from *Abies alba*, it is far less common on this species in Britain than formerly realised, due to its confusion with *Adelges nordmannianae*. Other hosts frequently attacked are *Abies grandis* and *A. procera*, particularly as a stem infestation. It is also to be found on *A. balsamea*, *A. amabilis* and *A. lasiocarpa* and can cause severe gouty swellings round buds and shoots (Plate 14) thus preventing all proper development of the shoots. Francke-Grosmann (1938) has described and figured many of these shoot deformities caused by *A. piceae*. When dense colonies occur on the stems of these North American *Abies* species the insects' feeding activities can upset the normal cell division in the cambium, so as to produce dense, brittle growth-rings known as "rotholz". Such a restriction to the transpiration-stream flow can have a serious effect on the subsequent growth of large trees (Doerksen & Mitchell, 1965). A heavy stem attack on *A. lasiocarpa* is known to kill trees in this way.

In Europe this adelgid is also known occasionally to infest stems of *Abies cilicica* and new buds of *A. fraseri*, *A. sibirica* and *A. lasiocarpa* var. *arizonica* (Francke-Grosmann, 1938).

E. Genus *PINEUS* Shimer 1869

The main morphological feature that separates *Pineus* from *Adelges* lies in the reduction of the abdominal spiracles in the former to four distinct pairs. The adult sistens are almost spherical when viewed dorsally, and frequently the heavily pigmented fused head and prothoracic shield can be seen without making a microslide preparation. Both alate forms (sexuparae and gallicolae) normally have no wax gland areas on the two extreme posterior (8th and 9th) abdominal segments. The sexupara nymphs have the head and prothoracic plates fused on either side of the mid line. The fundatrices do not have a fused head and prothoracic shield and the basic pattern of six rows of wax-gland areas is obscured by the general increase of wax gland areas over the dorsal surface.

In a similar manner to the genus *Adelges*, Börner and Heinze (1957) have divided *Pineus* up by separating off two of the North American species and putting them in the genus *Pineodes* Börner. These two species (*P. pinifolii* Fitch and *P. boycei* Annand) differ from the rest of the *Pineus* spp. by forming galls with scale-like chambers, and also in that the overwintering nymphs are very heavily chitinised with a distinct median row of abdominal wax glands.

All the holocyclic species have the two-yearly pentamorphic life cycle as tabulated in Chapter 2; the primary host being *Picea* and the secondary host *Pinus*. There are several species which appear to be anholocyclic either on *Picea* or *Pinus*. Recently Underwood and Balch (1964) have described a new anholocyclic species *Pineus abietinus* from *Abies amabilis* and *A. grandis* in North America, but the biology of this species is only partly known at present.

Galls are formed from developing shoots on *Picea* in the case of holocyclic species. They are somewhat looser than those of *Adelges* spp., having intercommunicating chambers. Other feeding sites on *Picea* are the needles, shoots and the stem – even of large trees. On *Pinus* there is a similar diversity of feeding site according to time of year and species.

Four species have been taken in Britain during the course of this work – *P. pini* (Macquart), *P. orientalis* (Dreyfus), *P. strobi* (Hartig) and *P. pineoides* (Cholodkovsky). The last species has not previously been recorded in Britain. The second species has not been recognised as a separate species in Britain until the course of this work. In order to separate these four species the apterae have had to be re-described.

KEY TO THE BRITISH SPECIES OF THE GENUS PINEUS
Adult Apterae Having an Ovipositor and Antennae with Three Segments

1 Prothoracic wax gland plates fused together and joined to the head forming a continuous shield   2
Prothoracic wax gland plates, numbering 20 or more, not fused together but remaining free. On foliage shoots of *Picea* .   .   .   .
.   .   .   (fundatrix) *orientalis* (Dreyfus)

2 The wax gland facets on the head-prothoracic shield are circular and of assorted diameters, they are loosely grouped but distinctly separate. If some of their margins do merge then they are not closely adpressed forming polygonal lumina. The groups of facets on the shield frequently occur in lighter pigmented areas. On *Pinus*   .
.   .   .   .   (sistens or progrediens) *pini* (Macquart) & *orientalis* (Dreyfus)

– The wax gland facets on the head-prothoracic shield occur in compact groups with their margins closely adpressed or fused so that many of the lumina are polygonal . . . 3

3 Posterior mesial glands on head-prothoracic shield usually having 10 to 20 or more facets. Anterior mesial gland facets on the head similar in size to those on the mesothorax. Meta and mesothoracic plates frequently pigmented and fused laterally. Pleural abdominal plates normally extend beyond segment two up to segment six. On *Pinus* . . (sistens or progrediens) *strobi* (Hartig)

– Posterior mesial glands on head-prothoracic shield absent or with only a few facets. Most anterior mesial gland facets on the head larger than those on the mesothorax. The facet lumina are frequently pigmented to the same density as the shield. The meta and mesothoracic glands occur mainly on separate plates. Pleural abdominal plates are not normally present beyond the second segment. On *Picea* . . (sistens or progrediens) *pineoides* (Chol.)

Alatae

1 Mesial wax gland plates on first abdominal segment normally fused to form one. Remaining abdominal mesial and pleural wax gland plates not represented on each segment. Those present usually have less than 8 facets. Sensorium on antennal segment V very large, more than half the segment length. Basal portion of sensorium about same length as terminal portion. Segment III with circular sensorium when seen ventrally . . . . . . *strobi* (Hartig)

– Mesial wax gland plates on first abdominal segment not fused. Remaining abdominal mesial plates well developed on anterior segments; the pleural plates being more prominent on the posterior segments. Sensorium on antennal segment V less than half the segment length. Basal portion of sensorium clearly longer than terminal portion. Segment III having a sensorium that is broad and angular . . . 2

2 Anterior marginal abdominal plates less than their own diameter apart. Pleural plates on the abdomen separate from the mesials and marginals at least up to segment 6. Marginal abdominal wax gland facets similar in size to the larger facets on the head. Fore-wing length 0.39 to 0.47 mm. . . . (sexupara) *pini* (Macquart) and *orientalis* (Dreyfus)

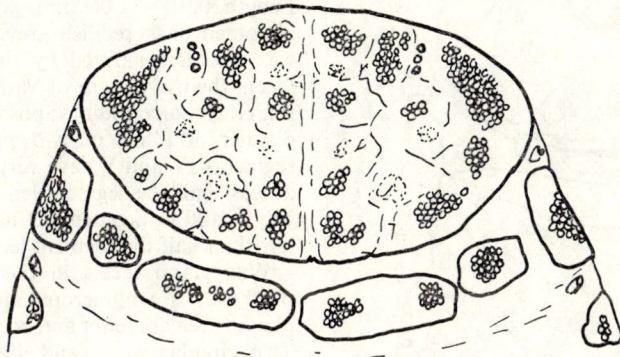

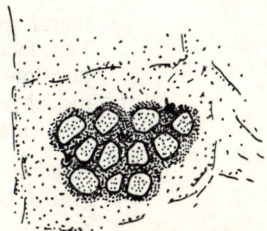

Fig. 24 *Pineus strobi:* head and prothoracic shield of adult sistens (above); detail of facet arrangement on the shield (below).

– Distance between the anterior marginal abdominal plates the same as, or more than, their diameter. Pleural plates on the abdomen fused or linked with the other plates on the same segment. Marginal wax gland facets on the abdomen smaller in diameter than the largest ones on the head. Fore-wing length 0.52 to 0.69 mm.
(gallicola) *orientalis* (Dreyfus)

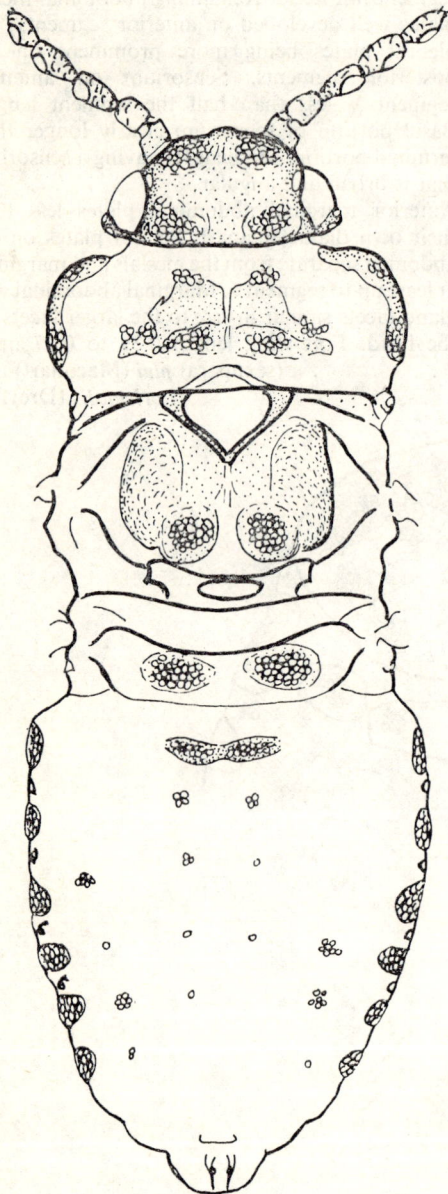

Fig. 25 Dorsal body surface of *Pineus strobi* (sexupara).

Pineus strobi (Hartig, 1837)

This species is well known as a pest in North America and Europe, feeding on the thin barked stems and branches of *Pinus strobus*. Descriptions and some biological details of this adelgid in Europe have been given by Börner (1908), Marchal (1913) and Pasek (1954). Originally this species probably came from North America where its host plants *Pinus strobus* and *Picea mariana* are native. The complete holocycle has not yet been recorded, but Marchal (1913) and Raske and Hodson (1964) describe the winged sexuparae and the subsequent production of female sexuales on *Picea mariana*.

MORPHOLOGICAL OBSERVATIONS
Adult Apterae
Length 0.75 – 0.9 mm, width 0.6 – 0.8 mm, body colour dark brown or dark red. The head-prothoracic shield is heavily chitinised. The wax gland facets have polygonal lumina, similar in type to *P. pineoides*, but are generally more numerous, occurring in large groups over the shield (see fig. 24). The eggs laid by the apterae measure between 0.26 and 0.28 mm long.

Sexuparae
Length 0.91 – 1.06 mm, width 0.33 – 0.45 mm, abdomen dark reddish grey. The anterior cephalic wax glands separated by the mid notal line. The facets themselves are of various sizes slightly paler than their surroundings; posterior cephalic facets are distinct and are in two patches. Both antennal segments IV and V have very large sensoria, usually at least half the segment length and extending ¾ the way round it. Segment III has a smaller sensorium, less than half the segment length (see fig. 26).

Wax gland facets in the marginal area of the prothorax are numerous; those in the mesial and pleural area are fewer and occur on a pigmented area. Two circular wax gland areas with distinct facets occur between the mesothoracic lobes; similar oval areas occur on the metathorax.

The forewings measure 1.3 mm long (average 10 specimens). The venation appears to be similar to the wings of *P. pini*. The hindwing media is very indistinct.

This species has smaller marginal wax gland plates than *P. pini* but the facets are equally well developed. Few facets occur in any of the mesial or pleural positions except on the first segment where the mesial pair are usually joined together forming one pigmented, faceted area (see fig. 25).

BIOLOGICAL OBSERVATIONS
Host Plants
Colonies with copious wax-wool secretions have been found on *Pinus strobus* stems in a number of localities throughout the South of England during

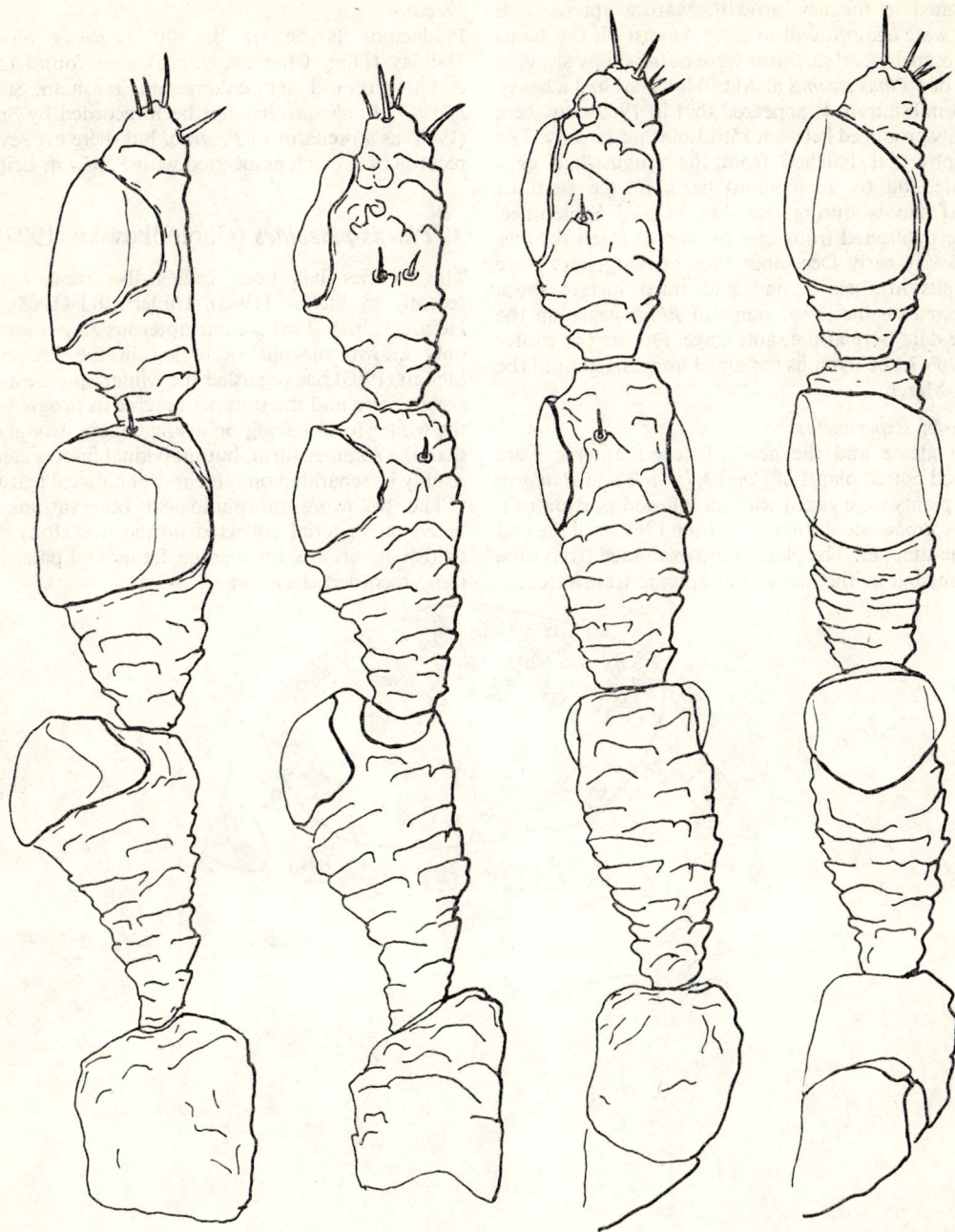

Fig. 26 Terminal antennal segments of *Pineus strobi* (sexupara)

this investigation. Colonies were also established on the stems of *Pinus peuce* (one of the European five-needles pines), at Bedgebury. During the winter months immature apterae were seen concealed under the papery scales of dwarf-shoots.

Observations on the Life Cycle
The earliest observation of mature apterae and egg clutches was 7th April 1965, which were present on the smooth bark of small branches at Alice Holt. By the end of May many of the hatching nymphs had

migrated to the new growth. Mature apterae and eggs were again found in early August on the stems and branches. Alate forms were obtained by sleeving part of a *Pinus strobus* at Alice Holt that had a heavy stem infestation. It appeared that in 1969 they were mainly produced between 15th June and 1st July. The nymphs that hatched from the August-laid eggs crawled on to uncolonised parts of the stem or dwarf shoots during late August and September. Insects collected from near the top of a felled *Pinus strobus* in early December 1966 at Bedgebury were nymphs in their second and third instar; those collected on the lower stems of *Pinus peuce* on the same date were at the same stage. During the winter of 1966/7, the nymphs remained unchanged until the 10th March.

Transfer Experiments

Both alatae and the newly hatched apterae were offered potted plants of *Pinus sylvestris* and *P. nigra*. The plants were caged with an infested portion of a *Pinus strobus* stem from early June 1969 until the end of the summer. The plants were examined from time to time but neither alatae nor apterae transferred.

Predator

Predacious larvae of the fly *Leucopis obscura* Haliday (Dipt. Chamaemyiidae) were found to be common on old stem colonies at Frensham, Surrey in 1967. *L. obscura* has not been recorded by Smith (1963) as a predator of *P. strobi* but there are several records of it being associated with *P. pini* in Britain.

Pineus pineoides (Cholodkovsky, 1907)

This species has been re-described and figured recently by Steffan (1963), Underwood (1963) and Heinze (1962). Two similar apterous stages are the only known morphs occurring in the life cycle. Steffan (1963) has regarded the winter apterae as the sistens form and the summer apterae as progrediens, the winter form having on average more wax-glands than the summer form, but individual insects cannot readily be separated on these morphological features.

The following morphological observations are based on material collected during this study. The British insects are, on average, larger and paler than those recorded elsewhere.

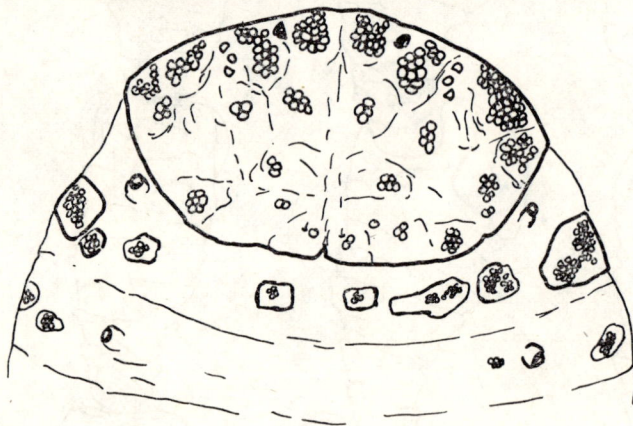

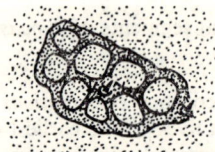

Fig. 27 *Pineus pineoides:* head and prothoracic shield of adult sistens (above); detail of facet arrangement on the shield (below).

Morphological Observations

Adult Apterae

Length 0.82 mm, width 0.61 mm, (average of 20 specimens), much smaller than the other British species in the genus. Dorsoventrally flattened, like an inverted saucer placed on another saucer, appearing circular when viewed from above. The chitinisation of the head prothoracic shield is weak, appearing a grey colour. Unchitinised parts yellowish-grey to very pale yellow. (Other authors describe the body colour as being a dark reddish brown). Gland facets on the head prothoracic shield are rounded and angular or nearly circular, distinctly bordered but not numerous (see fig. 27). The lumina of these facets are darker than the shield pigmentation. Wax gland areas are present on the meso- and meta-thorax, at least in the mesial, pleural and marginal areas, but the facets are, on average, smaller than those on the shield; there is frequently some lateral fusion of chitinised plates in these regions. Generally, only the marginal row of plates extends posteriorly beyond the second abdominal segment.

The eggs are few in number (up to 10) and are a pale orange colour.

Biological Observations

Host Plants – Distribution

The earliest record known of this insect's existence in Britain dates from January 1965 (specimens in the British Museum collection) from Cheshire where *P. pineoides* was infesting a Christmas tree plantation. *Picea abies* and *P. omorika* are the only two hosts attacked in Britain. So far this adelgid has been

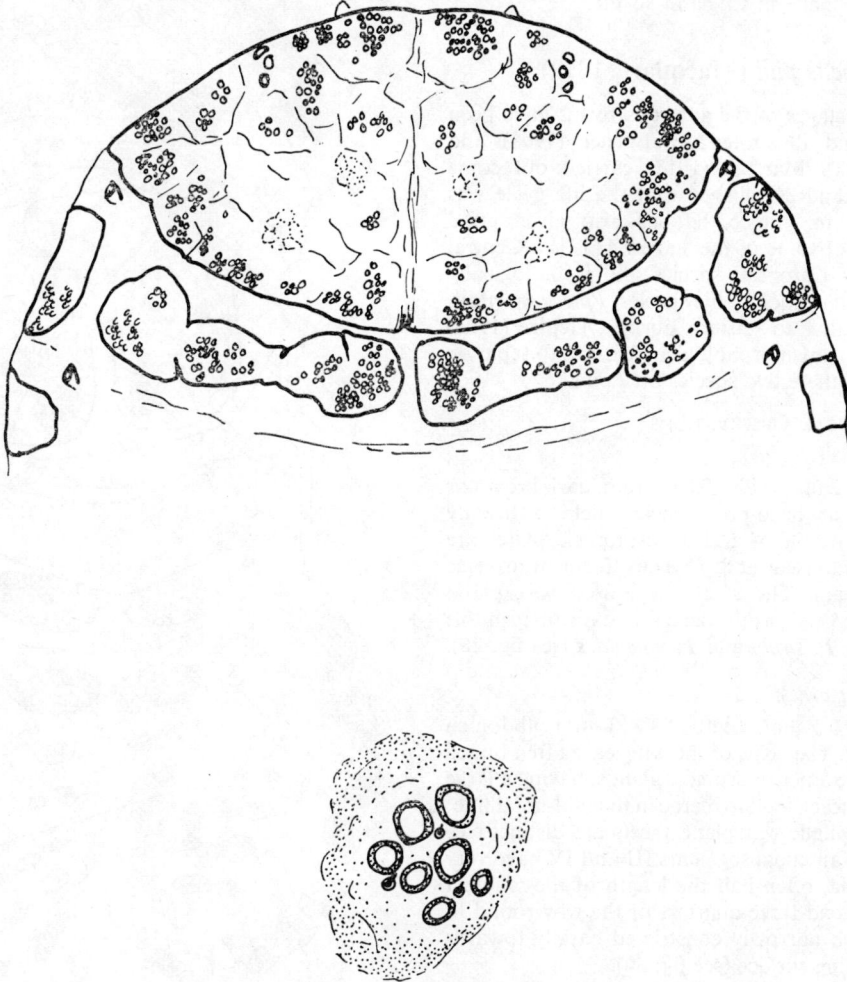

Fig. 28 Head and prothoracic shield of adult sistens of *Pineus pini* (above); detail of facet arrangement on the shield (below).

recorded from a few counties in England and is not common (Carter 1969).

This species has been found in Europe where its host plant, *Picea abies*, occurs as a native (Steffan 1963). More recently it has been found in Canada on *Picea rubens* (Underwood, 1963).

Observations on the Life Cycle

Adult apterae and eggs were collected in late May and again in July. Only immature apterae were to be found during the winter months.

Other workers have found from 2 to 4 generations a year. In New Brunswick, Underwood (1963) observed that the insects overwintered as second or third instars, that they developed into adults in May, and that their offspring became adult in July and then oviposited. This sequence compares with the few observations made in England so far.

Pineus pini (Macquart, 1819)

The various stages of the apterous forms have been described and illustrated by Börner (1908) and Heinze (1962). Marchal (1913) carried out some experiments and established that the life cycle was anholocyclic in France, he has also given good illustrations of some of the stages. This is a Central and Western European species and is biologically different from the holocyclic species, *Pineus orientalis* (Dreyfus), native to Eastern Europe. Heinze (1962) has not given any morphological features to separate the forms of these two species on *Pinus*.

MORPHOLOGICAL OBSERVATIONS

Adult Apterae (*Sistens*)

Length 1 – 1.2 mm, width 0.8 – 1 mm, dark brown or dark red. The head-prothoracic shield is heavily chitinised. The meso- and meta-thoracic plates are similarly chitinised and fuse to form transverse segmental bars. The small, separated wax gland facets on the head-prothoracic shield distinguish this species from *P. strobi* and *P. pineoides* (see fig. 28).

Alatae (*Sexuparae*)

Length 1 – 1.2 mm, width 0.8 – 1 mm, abdomen reddish grey. The veins of the wings are often tinged with red. The anterior cephalic glands, having a large number of facets, tend to merge in the mid-notal line; posterior cephalic wax gland facets are distinct (see fig. 29). The antennal segments III and IV have very large sensoria, often half the length of the segment and may extend three-quarters of the way round it. They are also narrowly constricted basally towards the articulating surface (see fig. 30).

Each mesial prothoracic wax gland area is joined to the adjacent pleural gland; marginal gland facets are numerous. Between the mesothoracic lobes is a

pair of posterior mesial wax glands with facets having distinct borders. A very large pair of mesial wax glands occur on the metathorax.

The forewings measure 1.4 to 1.75 mm long. Between the thorax and the stigma the costa is more or less parallel to the sub-costa. The medial and anal veins are slightly curved away from, and towards,

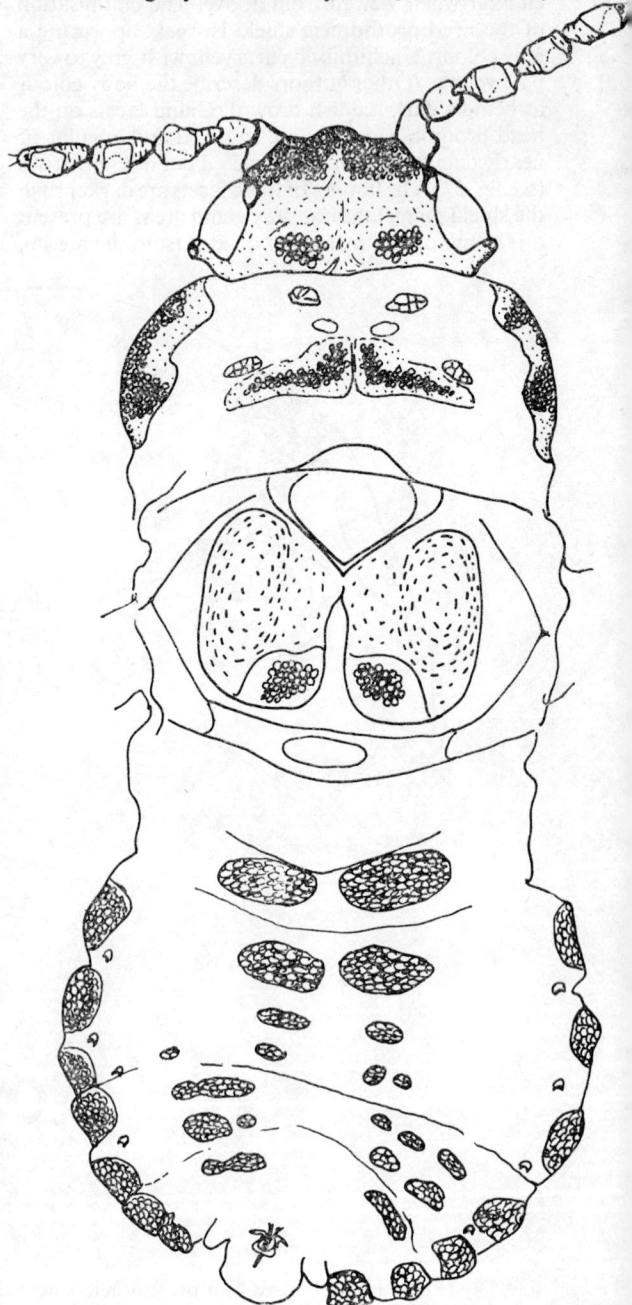

Fig. 29 Dorsal body surface of *Pineus pini* (sexupara).

the thorax respectively. The hind-wing media is indistinct, but it can be faintly seen curving away from the thorax (see fig. 31).

The wax gland facets on the abdominal plates have very clear borders and are larger than elsewhere on the body. Pairs of faceted mesial plates decrease in size from the anterior segments but on segments IV to VI they tend to fuse transversely with the mesials. Large oval marginal plates occur on segments I to III giving the abdomen a corrugated profile (see fig. 29).

BIOLOGICAL OBSERVATIONS

Host Plants

Largely restricted to *Pinus sylvestris*, but other two-needled pines are sometimes infested The winged sexuparae may settle on *Picea* after a period of active flight, this behaviour being a vestige of the holocycle. The species is widely distributed in the British Isles and in the rest of Europe where *Pinus sylvestris* occurs.

Observations on Life Cycle

A distinguishing feature of this species is the reduction of the number of forms in the life cycle. Colonies can always be found on the smaller stems and shoots of pine throughout the year. Those observed produced three or more generations of apterous forms. Large clusters of eggs were laid which resulted in con-

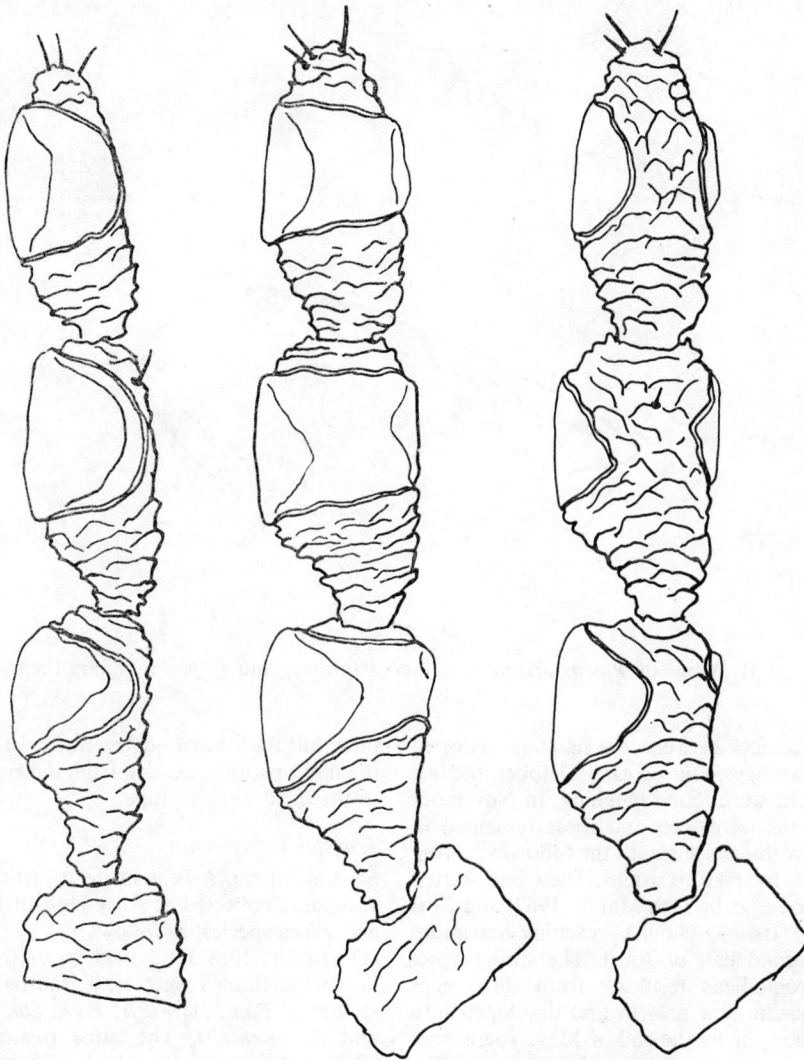

Fig. 30 Terminal antennal segments of *Pineus pini* (sexupara).

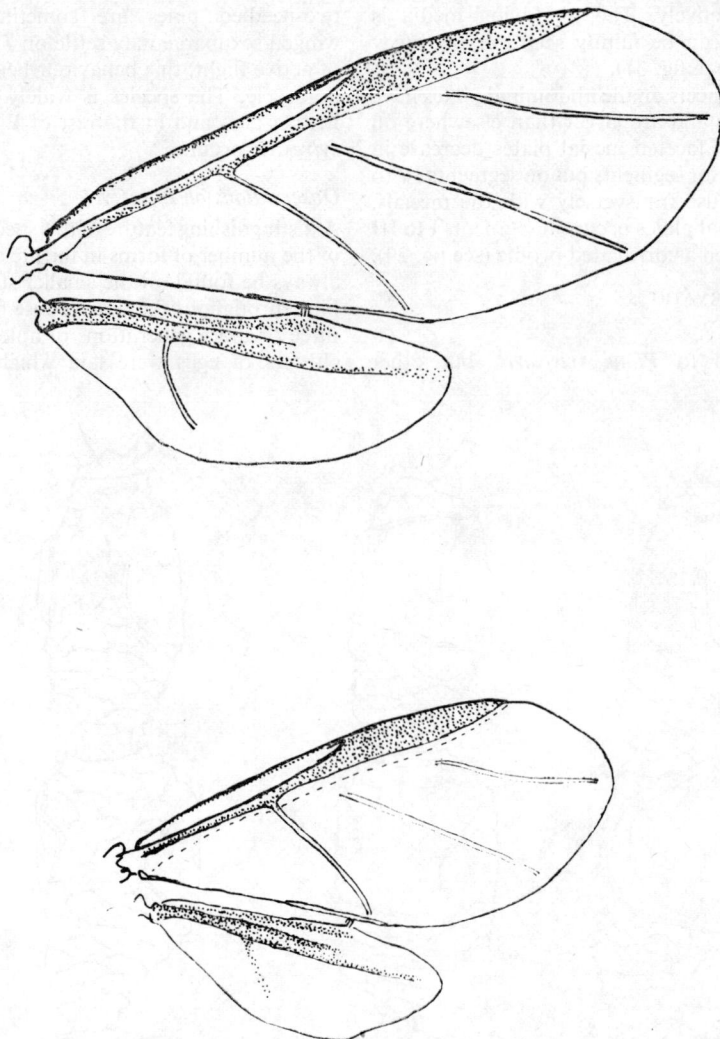

Fig. 31 Wings of *Pineus orientalis* gallicola (above) and *P. pini* sexupara (below).

siderable differences between the hatching dates of the first and last eggs. During early October, the last eggs of the year were found hatching. In November immature forms were seen and these remained in their second or third instar until the following spring when they became mature sistens. These had started to lay their first eggs by 11th March (1967) and 23rd March (1966), the egg clutches reaching maximum size in the second half of April. The spring generation of progrediens resulting from these eggs crawled on to the new growth and developed into two distinct morphs by the end of May. These were apterous progrediens and winged sexuparae. The apterae continued to feed and reproduce on the pine, but the sexuparae flew away. In 1966 and 1967 the flight period extended from the last week in May to the third week in June.

Transfer Experiments

Several attempts were made to artificially transfer sexuparae collected at Alice Holt to *Pinus sylvestris* and *Picea* species, as follows:

(1) On 2nd-10th June 1965 20 or more groups of flown individuals were each transferred to sleeved shoots of *Pinus sylvestris*, *Picea abies*, *P. sitchensis* and *P. orientalis*. The same treatment was also replicated on small potted plants. No progeny resulted.

(2) On 6th June 1966 50 sexuparae were transferred to a cage containing two-year old potted seedlings of *Picea abies*, *Picea sitchensis* and *P. orientalis*. Again no progeny resulted on the plants. Four days later it was noticed that some sexuparae had settled and died on the glass of the container, and that nymphs were under the shelter of their wings. On examination of microslides of the nymphs it was found that these were not, as might have been expected, sexuales. They had a distinctly elongate form and extremely long stylets looped round in their body. The dorsal surface had groups of gland facets on segmental plates arranged in the basic pattern (see fig. 7); the identity of this morph has not been established.

Pineus orientalis (Dreyfus, 1889)

This is the holocyclic species from which Marchal (1911a) considered *Pineus pini* to have originated. The holocycle occurs chiefly in Eastern Europe where the ranges of the two host plants, *Picea orientalis* and *Pinus sylvestris* overlap.

Marchal (1913) has described the sexual forms and Heinze (1962) has given a morphological description of the fundatrix.

MORPHOLOGICAL OBSERVATIONS

Adult Sistens

The generations found on *Pinus* are identical with *Pineus pini* and therefore cannot be separated.

Sexuales

These are orange and quite small when adult. The adult females measure between 0.44 – 0.61 mm, long and 0.27 to 0.38 mm wide. Pigmentation of their cuticle was very weak, having indistinct gland facets on the dorsal surface (see plate 17). The males are slender, being 0.57 mm long and 0.27 mm wide with long legs and very long III and IV antennal segments (see plate 16). They lack the dorsal wax gland facets present on the female.

Adult Gallicolae

Length 1.5 – 2.3 mm, width 0.7 – 0.95 mm. Colour deep red. This morph is larger but resembles the sexupara of *P. pini*, in having a similar distribution of dorsal wax glands and large distinct facets (see fig. 29). The antennal segments III and IV are of equal size and have large sensoria which are about half the length of the segments. The V antennal segment is longer than III and IV. The length of the sensorium on V, however, is less than half the length of the segment (see fig. 32).

The wax-gland plates on the prothorax are well distributed with facets. A pair of large mesial wax glands occur posteriorly between the mesothoracic lobes, and a similar pair occur on the metathorax.

The forewings are from 0.52 – 0.69 mm long. The costal and sub-costal veins are distinctly parallel up to the base of the stigma. The medial and anal veins are slightly curved away from, and towards the thorax respectively. The hind wings are between 0.34 – 0.43 mm long, and have a medial vein curving towards the distal tip (see fig. 31).

The facets of the marginal wax glands are smaller and more numerous than those of *P. pini*. They are also smaller in diameter than the largest facets on the head. The distance between the anterior marginal plates is normally greater than their diameter. In the

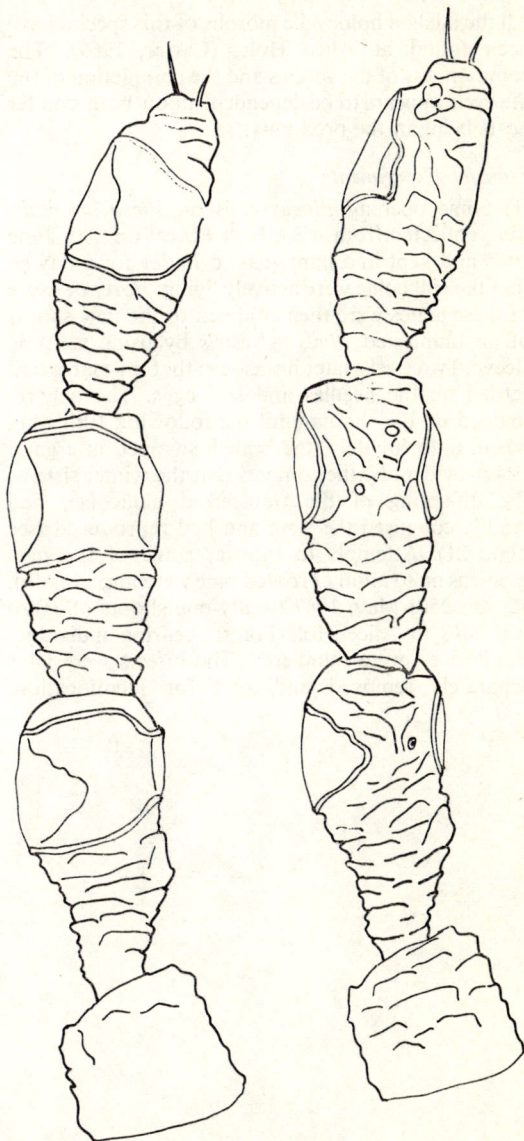

Fig. 32 Terminal antennal segments of *Pineus orientalis* (gallicola).

posterior mesial position the plates are always linked or joined to other plates on the same segment.

BIOLOGICAL OBSERVATIONS

Host Plants

This species has only been taken from the foliage and shoots of *Picea orientalis* at Bedgebury and Alice Holt. Adult sistens were obtained after becoming established on *Pinus sylvestris* following a successful transfer of gallicolae.

Observations on Life Cycle

All the adelgid holocyclic morphs of this species have been found at Alice Holt, (Carter, 1969). The commonness of the species and the completion of the life cycle appears to be dependent upon both conifer hosts being in the proximity.

Transfer Experiments

(1) Some opening *Pineus* galls on *Picea orientalis* were collected from Alice Holt Forest on 28th June 1965 and kept in a lamp-glass cylinder for a day so that the gallicolae were actively flying. Forty or more of these alates were then confined to the new shoots of an uninfested *Pinus sylvestris* by using a fabric sleeve. Two weeks later it was seen that the alatae had settled on the needles and laid eggs. The bag remained on the shoots until the following 10th May when, unfortunately, the branch snapped in a gale! It was evident by the contents that the winter sistens, the off-spring of the transferred gallicolae, had readily colonised the stem and had reproduced (see plate 21). Attempts to transfer some of the progrediens on to another rooted pine were unsuccessful.

(2) On 25th May, 1967 twenty-one shoots of *Picea orientalis* in Alice Holt Forest, bearing individual fundatrices, were numbered. The insects were later separately removed and kept for identification.

During the third week in June the galls were gathered. When the emerging gallicolae became active they were transferred to different species of pine. Twenty or more of the gallicolae were confined to new shoots of *Pinus sylvestris*, *P. contorta*, *P. nigra* and *P. strobus*. By the end of August it became apparent that only *P. sylvestris* had become colonised and that the insects on the other pines had died. The settled sistens nymphs seemed to prefer the axis of the needle-bases to other parts of the shoot. Development beyond this stage has not been studied.

CONCLUDING REMARKS

1. The winged stages of all the British adelgids can be identified by morphological features which are described here in detail. Individual insects of two species that differ in their life cycles, however, cannot readily be separated on morphological grounds.

2. The production of winged forms for each species appears to follow a seasonal pattern. Such information should be taken into account when chemical control operations are being planned.

3. The number of generations and morphs in the life cycles and the species of conifers attacked in Britain are, in some cases, different from those recorded as occurring elsewhere.

4. All known holocyclic species have a fundatrix stage that occurs on *Picea*. Further morphological studies on this stage are required for the proper identification of such specimens found feeding on *Picea* prior to gall formation.

5. Within the genus *Pineus* the morphology of the different species as described by a number of authors is very close. All nineteen species except one feed on either *Pinus* or *Picea*. It seems possible, therefore, that some of the world's described species may be synonymous.

Appendix 1

METHODS OF PREPARING SPECIMENS FOR MICROSCOPICAL EXAMINATION

Two techniques were used, one for the adult stages and another for first instars and the sexuales.

1. Technique for adult stages

The technique used was similar to the one for making permanent preparations of aphids given by Stroyan (1961).

(i) Material prior to preparation was stored in lactic alcohol. This had some clearing properties and prevented shrinkage to some extent.

(ii) The lactic alcohol was decanted off and the insects washed in 95% ethyl alcohol. Freshly preserved insects not stored in lactic alcohol were treated in the same way.

(iii) The insects were then sucked up into a small pipette which had a large orifice, and transferred to a small test tube and the alcohol level made up to about 2 ml. The tube was than put in a rack of a water bath and heated at approximately 85°C for ten minutes, after which the alcohol was carefully decanted off.

(iv) 1 ml of 10% potassium hydroxide solution was then poured into the tube and replaced in the water bath at the same temperature for precisely one minute. This time was found to be very important, as longer treatment removed pigmentation and made the exoskeleton soft. The potassium hydroxide maceration was halted by half filling the test tube with 95% ethyl alcohol.

(v) The insects were then tipped out with the solution into a solid watch glass and rinsed with at least two changes of 95% ethyl alcohol, using a pipette. Finally, all the alcohol was removed.

(vi) Then the insects were cleared in chloral phenol. Chloral phenol was made by adding equal parts by weight of chloral hydrate and phenol crystals and warming the mixture until it was a liquid. The liquid kept best in a dark bottle. Enough was poured on the insects in the watch glass to submerge them. The dish was then covered and put in an oven at 35°C for 12 to 48 hours.

(vii) Finally the insects were mounted in a gum chloral medium and placed in the oven at 35°C to dry the peripheral ring of gum – this usually took three to four days. Permanent slides were sealed by ringing with "Euparal".

Gum Chloral Mountant

This was made up with the following ingredients:

Gum arabic, clear colourless lumps	12 g
Distilled water	20 ml
Chloral hydrate crystals	20 g
Glacial acetic acid	5 ml
Glucose syrup 50%	5 ml

The gum arabic was first dissolved in the distilled water in an oven set at 35°C. When dissolved and still warm it was filtered through glass wool until all debris had been removed. The other ingredients were then added and the mixture replaced in the oven for a few days until all bubbles had dispersed. If the medium was too fluid it was kept in the oven for a few more days with the stopper off to evaporate.

2. Technique for first instars and sexuales

It was found that if the above technique was used for these small forms, odd individuals were frequently ruined by rapidly collapsing or drying out even if they were only momentarily bubbled up on to the sides of the test tube. A more gentle maceration technique was therefore developed.

(i) Insects, stored in lactic alcohol, 70% ethyl alcohol or freshly killed in 95%, were tipped into a solid watch glass and the liquid removed with a fine pipette. They were rinsed with 95% alcohol and then transferred in the alcohol, using a pipette, to small test tubes. The alcohol was then decanted off.

(ii) 1 ml of 75% lactic acid was then poured in the test tube and put in a hot water bath at about 80°C until the body contents were cleared. This required close inspection with a hand lens at intervals. Clearing times varied from 30 minutes to 2 hours according to age and pigmentation.

(iii) After tipping the insects and liquid into a solid watch glass they were rinsed with several changes of distilled water, using a fine pipette. During this rinsing process the insects became turgid, which was helpful in the final stages.

(iv) Final clearing was achieved by pipetting off the distilled water and pouring a small quantity of chloral phenol in the dish to submerge the insects. A lid was placed on the dish and put in the oven at 35°C for at least 24 hours.

(v) In order to examine the dorsal wax glands and pore arrangement it was found best to remove the ventral part, otherwise there was interference with the transmission of the light through the insect. When the specimen was put on the slide in a drop of gum chloral mountant it was turned on its side and, holding it with a fine needle, the legs and mouthparts were sliced off with a sharp hypodermic needle. The dorsal surface was then laid flat on the slide of the mountant before lowering the cover glass. They were then oven dried and rung with "Euparal" as the first technique.

Whilst carrying out both these techniques the collection number of the sample was kept alongside the specimens, this number being the reference to the data, was also the first thing written on the slide label. The host plant species, site of feeding, locality and date were subsequently added.

Appendix 2

WORLD CHECK LIST OF SPECIES WITHIN THE FAMILY ADELGIDAE

Genus and species	Main host plants	Distribution
ADELGES Vallot 1836		
ANISOPHLEBA Koch 1857		
CNAPHALODES Macquart 1843		
laricis Vallot 1836	Larix decidua	Throughout Europe. Also
coccineus (Ratzeburg 1843)	L. laricina	in N. America and
strobilobius (Kaltenbach 1843)	Picea abies	possibly oriental Asia.
geniculatus (Ratzeburg 1843)	P. sitchensis	
obtectus (Ratzeburg 1844)	P. rubens	
hamadryas (Koch 1857)	P. mariana	
atratus (Buckton 1883)		
lariceti (Altum 1889)		
consolidatus (Patch 1909)		
karamatsu Inouye 1953	Larix kaempferi	Japan
lapponicus Cholodkovsky 1889	Picea abies	Central Europe
praecox Cholodkovsky 1894		
tardus (Dreyfus 1888)	Picea abies	Continental Europe
niger (Solowiow 1924)		
affinis (Börner 1908)		
tardoides (Cholodkovsky 1911)	Larix russica	North East Europe
lapponicus (Börner 1932) nec (Cholodkovsky)		
japonicus (Monzen 1929)	Picea jezoensis	Japan
aenigmaticus (Annand 1928)	Larix laricina	North America
S. GILLETTEELA Börner 1930		
cooleyi (Gillette 1907)	Pseudotsuga menziesii	North America, Britain
coweni (Gillette 1907)	P. macrocarpa	
	Picea sitchensis	
	P. engelmanni	
	P. pungens	
S. CHOLODKOVSKYA Börner 1909		
viridana (Cholodkovsky 1896)	Larix kaempferi	Europe, Korea, Japan
	L. decidua	
	L. x eurolepis	
	L. russica	
	L. gmelini	
	L. gmelini var. olgensis	
viridula (Cholodkovsky 1902)	L. russica	Siberia
oregonensis Annand 1928	L. occidentalis	North America
	L. decidua	
S. APHRASTASIA Börner 1908		
pectinatae (Cholodkovsky 1888)	Abies sibirica	Northern Europe
	Picea abies	
coccineus (Cholodkovsky 1895) nec (Ratzeburg)		
ishiharai (Inouye 1953)	Abies sachalinensis	Japan
	A. sachalinensis var. mayriana	
	Picea glauca	
	P. glehnii	
	P. jezoensis	
	P. sitchensis	

Genus and species	Main host plants	Distribution
funitecta (Dreyfus 1888)	Tsuga canadensis	Western North America
tsugae Annand 1924	T. heterophylla	Formosa, Japan
	T. chinensis	
	T. sieboldii	
	Picea polita	
S. SACCHIPHANTES Curtis 1844		
viridis (Ratzeburg 1843)	Larix decidua	Europe
geniculatus (Ratzeburg 1843) in part	L. x eurolepis	
laricis (Hartig 1837) nec (Vallot)	Picea abies	
occidentalis (Cholodkovsky 1910)	P. sitchensis	
	P. orientalis	
abietis (Linnaeus 1758)	Picea abies	Europe, North America
gallarum abietis (Degeer 1773)	P. sitchensis	
alaeviridis (Solowiow 1924)	P. koyamai	
	P. jezoensis	
	P. glauca var. albertiana	
segregis Steffan 1961	Larix decidua	Continental Europe
laricifoliae (Fitch 1858)	Larix laricina	North America
	Picea glauca	
	P. abies	
lariciatus (Patch 1909)	Larix laricina	North America
	L. lyallii	
	Picea glauca	
	P. mariana	
	P. pungens	
karafutonis (Kono & Inouye 1938)	Picea jezoensis	Japan
kitamiensis Inouye 1963	Larix kaempferi	Japan
diversus Annand 1928	Larix laricina	North America
consolidatus (Patch 1909) in part	L. decidua	
S. DREYFUSIA Börner 1908		
nordmannianae (Eckstein 1890)	Abies nordmanniana	Europe, North America,
funitectus (Cholodkovsky 1907)	A. cilicia	Australasia
nuesslini (Börner 1908)	A. alba	
	Picea orientalis	
f. schneideri (Börner 1932)		
piceae (Ratzeburg 1843)	Abies alba	Europe
bouvieri (Cholodkovsky 1903)	A. nordmanniana	
	A. procera	
	A. balsamea	
	A. grandis	
	A. amabilis	
	A. lasiocarpa	
f. canadensis Merker & Eichhorn 1956	Abies balsamea	North America
prelli (Grosmann 1935)	Abies nordmanniana	Continental Europe
	A. cephalonica	
	Picea orientalis	
merkeri Eichhorn 1957	Abies alba	Europe
piceae f. aggressiva (Pschorn-Walcher & Zwölfer 1956)		
knucheli (Schneider-Orelli & Schneider 1954)	Abies pindrow	Himalayas
	Picea smithiana	
	Abies spectabilis	
? abietis-piceae (Stebbing 1903)		
? himalayensis (Stebbing 1910)		

Genus and species	Main host plants	Distribution
joshii (Schneider-Orelli & Schneider 1959)	Abies pindrow	Himalayas
todomatsui (Inouye 1945)	Abies sachalinensis var. mayriana	Japan

PINEUS Shimer 1869

    CHERMAPHIS Maskell 1884

    ANISOPHLEBA Koch 1857

Genus and species	Main host plants	Distribution

strobi (Hartig 1837)
   corticalis (Kaltenbach 1843)
   pinicorticis (Fitch 1856)
	Pinus strobus
	Picea mariana
		North America, Europe

orientalis (Dreyfus 1889)	Pinus sylvestris P. densiflora Picea orientalis P. thunbergii P. polita	Caucasus, Taurien, Europe, Japan, Korea
pini (Macquart 1819)	Pinus sylvestris P. ? mugo P. contorta P. nigra	Europe, Australia, New Zealand
laevis (Maskell 1885)	Pinus radiata P. thunbergii P. halepensis	California, Australasia
boerneri Annand 1928	Pinus radiata	North America, Formosa
havrylenkoi Blanchard 1944	Pinus sylvestris	Argentina
sylvestris Annand 1928	Pinus sylvestris P. radiata P. thunbergii	North America, Japan Madeira
coloradensis (Gillette 1907)	Pinus ponderosa P. contorta P. cembroides var. edulis	North America
floccus (Patch 1909)	Pinus strobus Picea rubens P. mariana	New England States
pineoides (Cholodkovsky 1903)	Picea abies P. omorika P. rubens	Europe, North America
patchae Börner 1957 similis (Gillette 1907) in part	Picea pungens	North America
cembrae (Cholodkovsky 1888) sibiricus (Cholodkovsky 1889)	Pinus cembra P. pumila P. koraiensis P. parviflora Picea glehnii	North Russia, Japan Siberia and Alps
abietinus Underwood & Balch 1964	Abies amabilis	North America
engelmanni Annand 1928	Picea engelmanni	North America
harukawai Inouye 1945	Pinus parviflora P. strobus	Japan
hosoyai Inouye 1953	Pinus koraiensis	China, Korea
kono-washiyai Inouye 1953	Picea abies	Japan
matsumurai Inouye 1953	Pinus densiflora P. thunbergii	Japan

S. PINEODES Börner 1957

Genus and species	Main host plants	Distribution
pinifoliae (Fitch 1858)	Pinus strobus P. monticola Picea mariana P. engelmanni	North America

Genus and species	*Main host plants*	*Distribution*
abieticolens (Packard 1869)	P. pungens	
pinicorticis (Hunter 1901) nec (Fitch)	P. sitchensis	
montanus (Gillette 1907)		
armiger Annand 1924		
boycei Annand 1928	Picea engelmanni	North America

NOTES ON THE LIST OF ADELGIDS

Generic names are printed in capital letters. The genera that are used by Börner and Heinze (1957), other than *Pineus* and *Adelges*, are included here only as sub-genera and are prefixed with the letter 'S'. Trivial names are printed in lower case letters and the usual practice of inserting the author's name in brackets when it is not in the original genus has been adopted. Synonyms are included where appropriate and are indented from the margin.

BIBLIOGRAPHY

ANNAND, P. N. (1928). A Contribution Towards a Monograph of the Adelginae of North America. *Stanford Univ. Un. Ser. Biol. Sci.* **6** (1): 1-46.

BLANCHARD, E. E. (1944) *Afidoideos argentinos. Acta zool. Lilloana* **2**: 52-57.

BÖRNER, C. (1908). Eine Monographische Studie über die Chermiden. *Arb. Biologischen Anstalt fur Land-und Forstwirtschaft.* **6** (2): 81-318.

BÖRNER, C. and HEINZE, K. (1957). Adelgidae (Chermesidae), Tannengalläuse. *Handb. Pfl-Krankh.* (Ed. Sorauer) **5**: 323-354.

BURDON, E. R. (1908a). Some critical observations on the European species of the genus Chermes. *J. econ. Biol.* **2**: 119-148.

BURDON, E. R. (1908b). The Spruce Gall and Larch Blight Diseases. *J. econ. Biol.* **2**: 1-13 and 64-66.

BUSBY, R. J. N. (1962). Species and Forms of the Silver Fir Adelgid in Scotland. *Scot. For.* **16**: 243-254.

BUSBY, R. J. N. (1964). A new Adelgid on *Abies grandis* causing compression wood. *Q. Jl For.* **58**: 160-162.

CAMERON, A. E. (1936). *Adelges cooleyi* Gillette (Hemiptera, Adelgidae) of the Douglas fir in Britain: Completion of its Life Cycle. *Ann. appl. Biol.* **23**: 585-605.

CARTER, C. I. (1969). Three species of adelgids (Homoptera, Adelgidae) new to Britain. *Entomologist's mon. Mag.* **105**: 167-169.

*CHOLODKOVSKY, N. A. (1896). Beitrage zu einer Monographie der Coniferen Läuse. *Horae. Soc. Entom. Ross.* **31**: 1-60.

CHRYSTAL, R. N. (1922). *The Douglas Fir Chermes (Chermes cooleyi).* Bull. For. Comm. Lond. **4**: 56 pp.

CHRYSTAL, R. N. (1924). The Pine Chermes in the Royal Gardens of Kew. *Gdnrs' Chron.* Dec. 6.: 388-389.

CHRYSTAL, R. N. (1925). The Genus Dreyfusia in Britain and its Relation to Silver fir. *Phil. Trans. R. Soc. B.* **214**: 29-61.

COTTIER, W. (1953). *Aphids of New Zealand.* Bull. N. Z. Dep. Scient. Ind. Res. **106**.

CROOKE, M. (1952). *Adelges Attacking Japanese and Hybrid Larches.* Forest Rec., Lond. **17**: 20 pp.

CROOKE, M. (1960). *Adelges cooleyi, An Insect Pest of Douglas Fir and Sitka spruce.* Leafl. For. Comm. **2**: 8 pp.

CUMMING, M. E. P. (1968). The Life History and Morphology of *Adelges lariciatus* (Homoptera: Phylloxeridae). *Can. Ent.* **100**: 113-126.

DOANE, C. C. (1961). The Taxonomy and Biology of *Pineus strobi* (Hartig) and *P. coloradensis* (Gillette) (Homoptera, Adelgidae). *Can. Ent.* **93**: 553-560.

DOERKSEN, A. H. & MITCHELL, R. G. (1965). Effects of the Balsam Woolly Aphid upon Wood Anatomy of some Western True Firs. *Forest Sci.* **11** (2): 181-188.

EASTOP, V. F. (1966). A Taxonomic Study of Australian Aphidoidea (Homoptera). *Aust. J. Zool.* **14**: 548.

ECKSTEIN, K. (1890). Zur Biologie der Gattung Chermes. *Zool. Anz.* **13**: 86-90.

EICHHORN, O. (1957). Eine neue Tannenlaus der Gattung Dreyfusia (*Dreyfusia merkeri* nov. spec.) *Z. angew. Zool.* **44**: 303-348.

EICHHORN, O. (1958). Morphologische und papier-chromatographische Untersuchungen zue Artentrennung in de Gattung Dreyfusia. *Z. angew. Ent.* **42**: 278-283.

FRANCKE-GROSMANN, H. (1938). Uber *Dreyfusia piceae* an ausländischen Tannenarten. *Tharandt. forstl. Jb.* **89**: 35-49.

GAUMONT, R. (1954). Le cycle du *Chermes viridanus* (Chol.) *C. r. hebd. Seanc. Acad. Sci., Paris.* **238**: 945-947.

HEINZE, K. (1962). Pflanzenschädliche Blattlausarten der Familien Lachnidae, Adelgidae und Phylloxeridae, eine systematisch-faunistische Studie. *Dt. ent. Z.* **9**: 143-227.

INOUYE, M. (1953). *Monographische Studie über die japanischen Koniferen-Gallenläuse.* Bull. Sapporo Brch Govt. Forest Exp. Stn. **15**: 91 pp.

MARCHAL, P. (1911a). La Spanadrie et l'obliteration de la reproduction sexuée chez les *Chermes. C. r. hebd. Seanc. Acad, Sci., Paris.* **153**: 299-302.

MARCHAL, P. (1911b). L'obliteration de la reproduction sexuée chez le *Chermes piceae* Ratz. *C. r. hebd. Seanc. Acad. Sci., Paris.* **153**: 603-604.

MARCHAL, P. (1913). Contribution a l'etude de la biologie des *Chermes. Ann. Sci. Nat. Zool.* **18**: 153-385.

PASEK, V. (1954). (Aphids attacking coniferous trees in Czechoslovakian forests). Vydavat. Slov. Akad. Vied, Bratislava, 319 pp. (in Czech).

*The original article has not been seen.

PHILIPTSCHENKO, I. (1916). (The Biological Species of Chermes and their Statistical Differentiation). (in French). *Zool. Vest.* **1**: 278-285.

PLUMB, G. H. (1953). The Formation and Development of the Norway Spruce Gall caused by *Adelges abietis* L. *Bull. Conn. agric. Exp. Stn.* **566**, 77 pp.

PSCHORN-WALCHER, H. & ZWÖLFER, H. (1956). Neure Untersuchungen über die Weisstannenläuse der Gattung *Dreyfusia* C. B. und ihren Vertilgerkreis. *Anz. f. Schadlingskd.* **29**: 116-122.

PSCHORN-WALCHER, H. & ZWÖLFER, H. (1958). Preliminary Investigations on the Dreyfusia Populations, Living on the Trunk of the Silver fir. *Z. angew. Ent.* **42**: 241-277

RASKE, A. G. & HODSON, A. C. (1964). The Development of *Pineus strobi* (Hartig) (Adelginae, Phylloxeridae) on White Pine and Black Spruce. *Can. Ent.* **96**: 599-616.

SHIMER, H. (1869). Notes on *Chermes pinicorticis* ("White Pine Louse"). *Trans. Am. ent. Soc.* **2**: 383-385.

SMITH, K. G. V. (1963). A short synopsis of British Chamaemyiidae. *Trans. Soc. Br. Ent.* **15**: 103-115.

SPEYER, E. R. (1919). Contributions to the Life History of the Larch Chermes, *Cnaphalodes strobilobius* Kalt *Ann. appl. Biol.* **6**: 171-182.

SPEYER, E. R. (1923). Researches on the Larch Chermes (*Cnaphalodes strobilobius* Kalt.), and their bearing upon the Evolution of the Chermisinae in general. *Phil. Trans. R. Soc. B.* **212**: 111-146.

STEFFAN, A. W. (1961). Die Stammes- und Siedlungsgeschichte des Artenkreises *Sacchiphantes viridis* (Adelgidae, Aphidoidea). *Zoologica, Stuttg.* **39**: 1-113.

STEFFAN, A. W. (1962). Zur Biologie und Okologie der europäischen *Sacchiphantes* – Arten in forstwirtschaftlicher Sicht (Adelgidae, Aphidoidea). *Z. angew. Ent.* **50**: 328-342.

STEFFAN, A. W. (1963). Zur Biologie, Morphologie und Systematik von *Pineus pineoides*. *Z. angew. Ent.* **52**: 286-297.

STEVEN, H. M. (1917). Chermisidae in relation to British Forestry. *Trans. R. Scott. arboric. Soc.* **31**: 131-155.

STROYAN, H. L. G. (1961). Identification of aphids living on *Citrus*. *Pl. Prot. Bull. F.A.O.* **9**: 45-65.

UNDERWOOD, G. R. (1963). *Pineus pineoides* (Cholodkovsky) (Homoptera: Adelgidae) on Red Spruce in New Brunswick and Nova Scotia. *Can. Ent.* **95**: 720-724.

UNDERWOOD, G. R. & BALCH, R. E. (1964). A New Species of Pineus (Homoptera: Adelgidae) on *Abies*. *Can. Ent.* **96**: 522-528.

VALLOT, M. (1836). Note sur l'Adelges laricis. *Nob. Mem. Acad. Sci. Arts Dijon* 1836: 224-227.

VARTY, I. W. (1956). *Adelges insects of Silver Fir.* Bull. For. Comm. Lond. **26**: 75 pp.

INDEX

References in **bold type** denote the main section to the subject. The appendices are not included.

Abies alba 30, 31, 32
  amabilis 32
  balsamea 32
  cilicica 30, 31
  fraseri 32
  gouty shoots on 32
  grandis 32
  lasiocarpa 32
  needle distortion of 30
  nordmanniana 30, 31
  procera 32
  sibirica 32
abietinus, Pineus 32
abietis, Adelges 1, **20**, 26
  separation from *A. viridis* 16, 20
  type of life cycle 4
Adelges **7**
  generic characters of 7, 32
  species with winged forms 8
  sub-genera within the genus 7
Adelgidae **3**
  key to genera 7
Aestivosistens 4
Alate females, morphology of 6
America, adelgids in 1
Anholocyclic life cycle 4
Aphidoidea 3
aphids, true 3
Aphrastasia 7
Apterous parthenogenetic females, morphology of 6

Balsam Woolly Aphid 31
Bottle-brush shoots 30
boycei, Pineus 32
British species, list of 7

Canada 31, 38
canadensis, forma 31
Cephalic glands 5, 6
Chermes, suppression of the name 7
Cholodkovskya 7, 12
cooleyi, Adelges 1, **22**

Dreyfusia 7

Eggs 3, 4, 12, 15, 30, 36, 37, 38
  hatching synchronised with bud burst 26, 32
  long period of oviposition and hatching 30, 40
  of fundatrices 31
Europe 1, 38, 39
  Silver Fir Adelgids in 31
Exsules 3

Feeding sites 7
First instars, morphology of 5
France 1, 16, 38

Fundatrix 3, 42
  activity in spring of 30

Gallicolae 3, 11, 20, 24, 29, 41
Galls 3
  atypical forms 26
  elongate 25
  formation of 3, 20
  in clusters 20
  of *Adelges* spp. 8, 12, 20, 26, 30
  of *Pineus* spp. 32, 42
  on *Picea abies* 12, 20, 21
  on *Picea orientalis* 20, 30, 42
  on *Picea sitchensis* 12, 25, 26
  pineapple type 16
  time of opening 21
  with scale like chambers 32
Gamic morphs, loss of 4
Germany 1
Gilletteela 7
grandis, forma 31

Hibernation 7
Hiemosistens 4
Holocyclic life cycle 3
Honeydew 12

Identification, by biological features 7
  by chromatography 1
  by morphometric techniques 7

Japan, adelgids in 1, 12

Key to, American species 1
  British *Pineus* spp. using apterae 32
  British species using alatae 8, 33
  European species using first instars 1
  Genera 7

Larch, adelgids on 1
  needle distortion 12
  needle kinking 20
  long shoots, adelgids on 15, 16
  stems, adelgids on 14, 15
laricis, Adelges 4, **9**
Larix decidua 12, 14, 20
  gmelini 14
  kaempferi 14, 16
  laricina 12
  occidentalis 12
  russica 12, 14
  x eurolepis 12, 14, 20
  x pendula 12
Leucopis obscura 36
Life cycles 1, **3**
  anholocyclic 4
  holocyclic 3, 4

50

Printed in England for Her Majesty's Stationery Office by St Clements Fosh & Cross Ltd.   Dd 501973 K